# Accidental Death

*a novel by*

# *Iris V. Penn*

To: Dear Maureen,
With love and Best Wishes,
Iris.

©Iris V. Penn 2004   ISBN 0 906522 29 3

# Chapter One

Accidental Death.

The words were indelibly printed on my mind. But was it? Could there have been a conspiracy and Ken happened to have been in the way?

I tried to pull myself together, for after all, I reasoned to myself, that sort of thing would never happen in the sombre world of British Horse Racing.

Again and again I went over the events leading up to Ken's untimely death, from the day I had started out so happily to meet him at Kempton. It was one of those rare, glorious days in October, and on the drive home I remember glancing at the dashboard clock, before switching on the radio. I felt the usual ripple of excitement building up inside me, as the Commentator announced "...racing from Kempton". I knew every first, second and third up until the last race by heart. I listened intently for the two horses I had backed that afternoon. Good prices they were too. 6/1 and 11/2. I felt my hands beginning to sweat up on the steering wheel, as his voice went on "…Catterick 1:30 – I knew only too well how temperamental some horses can be before a race, and my pulse was shooting up by the second. It was a toss-up whether I wanted him to get a move on and let us have the result of the 4:30 or to prolong the agony and live in hopes.

This wasn't all due to the fact that I had had ten quid on Music Adored myself, and wanted to make it a hat-trick on the day. On no, not by a long shot. It was because it was Ken's horse. That was why 1 so desperately wanted it to win, for his sake.

My fingers were tapping out a tune "come on… come on" I was willing it. Every punter knows the feeling.

The radio crackled on "…4:30 Music Adored 11/2. Eleven-to-two, I couldn't believe it. A fantastic price. I would have been more than delighted for it to have been returned at 7/2.

I wondered how Ken was feeling – on cloud nine I imagined. The only disappointment being that he was unable to get to Catterick to lead it in.

He had a couple of appointments in Town in the morning that he was unable to break, and could only make it to Kempton in time for the second race.

I normally saved all my annual leave to take in separate days throughout the season so that I could go to the weekday meetings I liked best. Sandown, Kempton and all round courses where I could follow the horses with my glasses.

I have a severe hearing problem and therefore am unable to hear the commentary. Newmarket and

Doncaster are absolutely beautiful but not good for me, on a straight course I can only pick up the jockeys' hats at the two furlong marker and by the time I've sorted them all out, it's all over bar the shouting. We had a good relationship, Ken and I. We not only shared the same interests of National Hunt Racing and music, we also shared the same office. The place wouldn't have be the same to me, without seeing his dark unruly mop of hair bent over his desk, or seeing his tall, lean figure striding about. He wasn't exactly handsome but there was something about his craggy face and kind heart that melted my heart.

What he saw in a five-foot three, little overweight, mousey blond, who was more than a little scatty, always puzzled me, but there was a time when he asked me to marry him being such an honourable gentleman but with my widow's pension and a job that was very interesting, and not too demanding, I wasn't too keen to give up my independence, and the extra domestic duties that go with marriage, didn't appeal to me in the slightest. So I put it to him, as gently as possible, that if it was o.k. by him, I was quite happy continuing as we were, and I felt he was quite relieved. He had spent his whole adult life working to achieve his ambition of becoming Underwriter of Livestock at Lloyds, and had left it a bit late at forty to change the scene.

We had left before the last race at Kempton, and before the result from Catterick, as Ken had wanted to get back to the office in order to sign a few letters that were to go out that night, and we had decided to stop off for a meal later. It had become the usual thing for us to splash out on a meal if one of us had had a good winning day. Otherwise, we usually took some fish and chips home from Chris's Plaice, or a Chinese take-away from Pang's. Not bad, not bad at all, but not up to the standard of the little pub in Essex we had intended to go to where the food was first-class and the price comparatively reasonable. There we could hold a conversation without having to shout above the sound of music.

I guessed we would be having champagne that evening to celebrate Music Adored's win, as this was something really special. I imagine Ken must have been feeling very thrilled, particularly as Music Adored was one of his favourite horses. He loved them all really, but there was no doubt in my mind that Music Adored was his favourite.

I could see the rear lights of Ken's BMW two cars ahead of me, just rounding the high curve on the flyover. In my wing-mirror I noticed an H.G.V. coming up on the outside lane. I then lost sight of the BMW for some time, and eventually picked it up to see it swerving all over the road? as though Ken was drunk, or a tyre had burst. It was completely out of control. I tried to imagine what had happened. Perhaps Ken had collapsed... he certainly didn't seem to be making any effort to get the car straightened up. Or was it that the steering had gone?

Eventually, the inevitable happened. It hit the barrier, turned a couple of somersaults, and came to a crashing halt upside down. I pulled over onto the hard shoulder, together with other drivers who had miraculously missed all the rolling and swerving. We all rushed over to the overturned car, and fought like mad to try and release Ken from his seat belt, as the trickle of petrol began to increase in volume, but his right leg was badly trapped and there was no way we were going to be able to get him out, it was a job for the Firemen, who soon arrived, together with the Police and Ambulance. I walked back to the Escort amid the barrage of flashing blue lights and followed the Ambulance to the Hospital.

In Casualty they were ready and waiting, and immediately went into action.

I was shown into a waiting room, and given a hot sweet cup of tea. The wait seemed endless. I could smell all the familiar smells associated with hospitals, and see the white coats of the Doctors flapping around their knees as they walked too and fro, and I remember thinking how dignified the nurses were in the starched aprons and frilly caps, but it was like a dream or rather a nightmare.

It was absolutely unbelievable that one could be so happy one moment and completely devastated the next.

Eventually the nice young Doctor looking absolutely exhausted with his fair hair flopping over his face, and shoulders sagging, came to see me. He told me, as kindly as he possibly could, that although they did everything to try and get his heart going again, the injuries were so severe, that he had lost the battle for life just a few minutes previously.

This was no dream or nightmare – it was for real, all too real.

I ambled around for the next couple of days in a trance-like state. Nothing, absolutely nothing seemed to register.

There had to be a post-mortem and inquest of course, and Steve Allen, Ken's Solicitor, who was also his friend, was attending to all the details.

The body was finally released and Steve made all the necessary funeral arrangements, which I sat through like a zombie, completely unable to take in or believe what was happening.

Still in a state of shock, all I could think of, was that life was going to be pretty dull without Ken. He was such a necessary person, so alive, so good, not vain or boastful of the success he had made of life, just a nice guy, one of the best.

Leadenhall Street on a wet Monday morning looked just about as dreary as I felt.

I knew this was going to be the hardest part of all. Walking into the office and seeing Ken's empty desk. No cheery greeting, no "good morning Val, what have we got exciting in the post today?" The work had piled up, and I knew there was no alternative but to make a start on tackling it, but it was not easy to raise the enthusiasm needed to do so.

The miserable day wore on, and I seemed to be getting nowhere fast. I had constant interruptions all morning, the telephone had buzzed incessantly, and out of the corner of my eye I could see the light on it flashing away again.

I breathed a deep sigh and flicked the switch. Joyce at the other end apologized for troubling me again, but said that she thought it was important as it was Steve Allen of Allen & Moore the Solicitors on the line.

Would I take the call?

## Chapter One

Yes thank you Joyce" I answered a little wearily "Hello Steve, what can I do for you?" His voice sounded concerned as he replied "Valerie, how are you?" "Fair" I replied "bearing up". His voice softened "Look Val" he went on "do you think you could possibly pop in and see me one day this week or better still could we have lunch together?" I hesitated, wondering how I was going to fit it in with all the additional work at the moment, and then realised that Steve would not have asked me if it was something unimportant he wanted to see me about "Of course, Steve, I should like to" "Would Wednesday suit you, say about one at the Palmerston? " I flicked through my diary. "Fine" I answered "I shall look forward to seeing you Steve, bye for now".

The day progressed in the same dead-pan way, and I finally joined the mad rush home from Liverpool Street, not even bothering to try and dodge the stampede of feet, umbrellas and brief cases.

Race horses didn't know how lucky they were with their travel arrangements, owners and trainers would never allow them to travel in trains packed like sardines as we had to.

Digging a few people with my elbows, I managed to turn to the back page of The Standard to see what had won at Warwick. Nothing very exciting there, or was it that I couldn't raise any enthusiasm?

Was it always going to be like this, I thought with despair, or in time would life be worth living again?

The front page of the paper was full of depressing world news, and the all too familiar muggings of old people.

# Chapter Two

The following Wednesday I found that Steve had reserved a table at the Palmerston, and I wished that I felt more in the mood for enjoying the fare patronised by the cream of the City.

After ordering and going through the few normal preliminaries, Steve came straight to the point.

"You know, of course, Valerie, that Ken was, like myself, a Barnardo Boy?"

I didn't and it shook me, but when I came to think about it, Ken had said very little about his early days. He had always given me the impression that he had had quite a happy childhood, doing all the things boys normally do.

Steve's voice broke into my thoughts "...so quite naturally he has left the bulk of his Estate to the Charity who had looked after him so well when he was young – Barnardos."

I could well understand that, for it was typical of

what I would expect of Ken, who had always showed his appreciation for the smallest act of kindness bestowed upon him.

But if that news had shaken me, Steve's next surprise really bowled me over…

"But he left his most precious possession to you"… he went on "his beloved horse Music Adored, together with sufficient funds to pay for training fees, and all other expenses."

To say I was staggered was the understatement of the year. I went through so many emotional tremors at once, I didn't know which one hit me first.

The first, I think, was that it seemed that Ken had really loved me, which surely he must have done to have left me what he valued most in life, the horse he had named after his life-long love of music. Then sadness, that now I would never be able to let him know just how much he meant to me. And after that, *joy*, such unbelievable and overwhelming joy, that I had actually become an owner.

It had always been my dream to have a horse of my own, and I could never, ever, really begin to imagine what it would be like, but for Music Adored to be mine, the horse that Ken said he reckoned was going to be the greatest since Arkle was a little too much to take in in one go.

Anyway, when all the joy, sadness and heart-throbbing began to settle, and I started to come back down to earth, I realised I hadn't asked Steve all the important questions. He had been so patient while I had been going through this emotional upheaval, quietly getting on with his drink and obviously realising that it was a lot for me to take in.

"What was the result of the inquest Steve?" I felt I couldn't go to the Coroner's Court and go through all the traumatic details again. Somewhere along the line I had just taken it for granted that the car must have been faulty, or Ken had suddenly been taken ill.

"Accidental death Val, what we had anticipated."

"Accidental death? Oh come on Steve, we all know there had been an accident, but what had caused it – had Ken been taken ill?"

"Not according to the Autopsy, apart from the injuries he received, he had been in perfect health." "Then it *was* the car, what was it, a burst tyre or steering?" "Not according to the Insurance Company, their man has given it a very thorough inspection and could find nothing to indicate a prior fault."

He put his glass on the table and beckoned the waiter for a refill. "Will you have another Val?" "No thanks Steve, I have a lot of work on hand this afternoon, and must keep a clear head."

I was getting very agitated, and thinking back to the night at Stratford, when Ken became so distraught about the men we passed in the car park, I was beginning to believe somehow a cover-up was taking over.

"Look Steve, let's look at this in a logical way, the accident must have been caused by something – what did they say at the inquest?" "As I said before, it was brought in as accidental death, and the only thing they could put it down to was that Ken must have probably dozed off for a few seconds or lost concentration after hearing the result of the 4:30 at Catterick. People do doze off when driving you know, it happens all the time." "Maybe, I said in disbelief, but not with people like Ken." He was an expert driver, and very conscious of his responsibilities towards other people. I would never, ever, believe or accept the fact that Ken had died through his own carelessness.

The lunch was so good, and so beautifully presented I found that, without realising it, I was actually eating, and what's more, enjoying it. Something I had not done for the past few weeks. The food, together with the wine, made me feel much more relaxed and less disorientated, and I decided there and then to begin to sort things out in a rational way later in the day when I was home and quiet, and could go through every detail and try and make some sense of the whole bewildering affair. Steve had said that Ken had recently updated his Will, which was strange for a man of his age, something or somebody was obviously worrying him. As we walked down the steps of the Palmerston into Bishopsgate, Steve said he would be in touch, as there were papers to be signed and decisions to be made.

I had already decided that I would continue to race in Ken's colours, blue and claret as he was a West Ham fan.

Ken had, apparently, stipulated very strongly in the Will that he would like Music Adored to remain with her present trainer, Simon Galloway, who although young, had had a successful record so far, and certainly believed in modern techniques, and that the horse should never be returned to the Trainer who bred her, Amanda Neale-Adamson. This apparently was the main reason for the update of the Will.

Arriving home, after the usual scrum from Liverpool Street, and the depressing weather, I put on all the heating and restful lamps, made myself a coffee, thankful that I didn't have to prepare food, which I had been in the habit of not eating, snuggled down into the comfy sofa and tried and tried to figure out how a perfectly fit man, with a perfectly good car, could just pile up like that.

Ever since the accident I had been going over in my mind those last few terrible minutes again and again, only to come up each time with the same answer, nothing.

# Chapter Three

The Anglo Catholic Church I attend was looking more lovely than ever on the Sunday.

We had celebrated our Patronial Festival that week, and the devoted ladies had cleaned the Church, polished the brass and arranged the flowers to perfection.

I dipped my finger in the Holy Water and crossed my forehead as I entered, and made my way to my usual pew down the front in anticipation of our guest Preacher not having a beard, and that I would, therefore, be able to lip-read at least a little of what he was preaching.

It had never occurred to me that I was particularly religious, but I had always thought that High Mass was one of the most moving and beautiful things in the world;, and I had always enjoyed meeting, once a week at least, the people I had loved and respected all my life. It was all so familiar to me, and yet it felt so strange.

The Priest gracefully gliding towards the Altar in his

colourful vestments, the sacristan and servers busily carrying out their various duties, remembering to genuflect as they passed from one side of the chancel to the other, the sung responses, and above all the the overpowering small of incense.

I must have been dreaming. I came to with a start as the man beside me extended his hand "Peace Be With You". The one thing I wanted more than anything else, but was unable to find. I was confused, unsettled, and felt like I was on the outside looking in.

When the Mass was over and the congregation had gone over to the hall for coffee and a gossip about what had taken place the previous week, and what was about to happen next, I went over to the Shrine of Our Lady and lit a candle for Ken.

I knelt before the Statue of the Holy Mother, and desperately wanted to say a special prayer. Somehow, it just wouldn't come. I couldn't find the right words, but as I knelt there I gradually began to feel serenity creeping back into my body. Peace was beginning to be with me again. My thoughts were starting to sort themselves out, and find their way into the right slots.

Although the right prayer wouldn't come to my lips, I felt that the Holy Mother was getting the message, and that she would pray for me.

Then suddenly, for the first time, it came into my mind that perhaps what everyone was thinking of as an accident, had not been an accident at all.

Could it possibly be that someone had deliberately tried to put the BMW, and its driver out of action. Not Ken of course, for everybody who knew him, liked him.

## Chapter Three

I wondered could they have mistaken the car and the driver for somebody else that they wished to dispose of.

The thoughts went round and round in my head like a cartwheel, and the more I thought about it, the more certain I felt that there had to be a reason for the car to go out of control the way it did. I would never, ever, he able to accept the view of the Coroner that Ken's death was accidental, and kneeling there before the Holy Mother, I felt I was being given an inner strength to somehow go out and try to get to the bottom of this mystery.

Above all, I felt I owed it to Ken in return for all that he had done for me. He had been great to me when alive, and now, although my dearest wish would be to have had him alive and well, he had left me what he prized most, his beloved Music Adored.

For this great honour I was determined to do my best to prove that it was not Ken's fault.

I knew it was not just instinct pressing me on, for my whole body was screaming out that there was something very phoney about the whole affair.

I walked out into the sunshine of the beautiful autumn day, and it was like coming out of a thick fog.

I had made up my mind to do something about the situation, and already I felt better, much, much better than when I had gone into the Church that morning.

# Chapter Four

The first thing to do was to get the car checked over. Dick, my late husband, had been in the motor trade all his life, and often used to test-drive cars on the tracks for the brake and tyre companies, and knew most of the mechanics quite well.

One of them, a chap called Bob, had written to me when Dick died, saying he would always pop over and give my car a service for me. I hadn't taken him up on that, but I still had all the letters and cards I had received at the time, and thought it worth a stamp to see if he was still living at the same address. He was, said he had retired, and was delighted to have the opportunity to get his hands on something other than the routine jobs he was now doing for his friends and relations. The Insurance company gave their permission for an independent inspection, and I faxed over the details to Bob that night, where it was being kept.

He came back to me the following day suggesting we

met for a drink at a little pub in the City just off Broad Street, where we would be able to talk without interruption.

It was good to see this cheery little chap again, he was so genuine and unassuming, and by all accounts had been an excellent mechanic to one of the top racing drivers.

I put Bob into the picture as to what had happened, and what had been said at the inquest.

"I've gone over that BMW with a fine-tooth comb, Valerie, and apart from the bumps it had taken, I found it perfectly sound. Nothing, absolutely nothing, was I able to fault that would have been responsible for the crash."

This was not easy for me to accept. Somewhere at the back of my mind I had been so sure that the Insurance man had missed something. After all, he was seeing cars every day of the week, surely one was very much like another to him, but if Bob said there was nothing wrong with it, then I had to accept his word for it. I had enjoyed seeing him again, he really was a dear, positively refusing to accept any payment or expenses for the past two days, saying how much he had respected and liked Dick when he knew him in the old days, and how Dick would always go out of his way to get spare parts for Bob, that were a bit dodgy to get hold of at that time.

He said the boys used to call Dick the King-pin King.

All this served to make me even more determined to investigate further, but what the next step to take was going to be I had no idea. The only thing I could do was to play it by ear.

Perhaps Steve would have some suggestions to make, I thought I would have to give him a ring.

I had had the most uneasy feeling since the night before the accident. Ken had been sent a couple of tickets for the "first night" at the Theatre Royal, Stratford, East London of a play called "Stiff Options". Ken was Vice President of the Waltham Forest Brass Band, and they were going to be there to play to greet the celebrities who had been invited to see the play.

The theatre had been completely redecorated and refurnished, and the Management had opened their doors to the Gentlemen of the Press and the World of Showbiz.

The streets of East London were awash with Rolls, Porches and Jensens, and the East Londoners had turned out in force in their good-hearted way to welcome them.

The Waltham Forest Band, Ken's band, were looking very smart in their blue and black uniforms, and were seated in the paved Gerry Raffles Square outside the theatre, ready to strike up with songs from the shows, as the guests and press began to arrive. As the crowds began to thicken, the atmosphere started to warm up, and Simon the Musical Director, was getting well into his stride, urging the Band to give it everything they'd got. It wasn't long before the crowds lining the grass banks, and flanking the Square, were singing the good old tunes, and it only took "If you knew Susie" to get them dancing, the youngsters allowing the old 'uns to show them the way, and in no time at all there was dancing all round the Square. Every now and again there was a pause to send up a rousing cheer, as one of their favourite stars arrived, but most of the celebrities were only inside long enough to grab and drink and some of the tasty little snacks before they returned to the outside fun. A call then came for them to take their seats as the curtain was due to go up in five minutes.

The West End had come to the East End, and it turned out to be the best knees-up since the Royal Wedding. Like everyone else, that night, Ken was having problems finding a parking space, and he said "come on Val, there's only one thing for it, lets make for the multi-storey car park over the shopping centre." We were so happy, the festive mood had captured us, and hand in hand, laughing together at such silly little things, we were making our way back to the theatre.

As we reached level 1 of the car-park there were two men who appeared to be engaged in a very serious conversation, one of whom I recognized as a man called "Chopper". He was currently head of a bunch of crooks and had got his nick-name by having three of his fingers of his left hand chopped off by a rival gang, and whose picture appeared with monotonous regularity in the papers. He had a stall in Walthamstow High Street, after his wife left and divorced him because he lost the licence of the pub her father had set them up in when they married, and now had to get the money for the life he had become accustomed to by running a protection racket round the East London markets. I could only see the back of the man he was talking to, but Ken must have known who it was as his mood appeared to change from that moment, and whoever it was, it kept his thoughts pretty well occupied all evening. He tried hard not to let it show, but I was very sensitive as far as he was concerned, and I was becoming more and more worried as the night wore on.

It was the most incredible coincidence that I should have been there, sitting in the theatre enjoying a play by John Flannagan and Andrew McCulloch, whose T.V. work included, among many others, The Sweeney and The Heavy Mob. Never in my wildest dreams, could I have visualised that my own life was soon to unfold for real, a similarity to what these two gifted gentlemen wrote.

The drive home was almost in silence. "What's up Ken, it's not like you not to enjoy a good joke, and the play was very funny?" "Of course I enjoyed it, it's been a fantastic evening." and I remember as he just brushed my cheek and said "Goodnight Val, God Bless" I replied "Goodnight Ken, I hope you can make it to Kempton tomorrow, I'll look out for you." If only I could take those words back now.

# Chapter Five

In the meantime Wetherby's were being most helpful with all the information a new owner has to know, and Music Adored was re-registered in my name, and the blue and claret colours we had had so much fun in dreaming up when Ken had bought his first horse. The boys and girls at the office had had a whip round on my Birthday and came up with a smart new blanket for Music Adored, the very latest, with my initials beautifully embroidered in the corner. I was completely overwhelmed by their kindness and thoughtfulness, and had to make my way quickly over to the table to get busy with the little savouries, cakes and wine that we made birthdays an excuse for indulging in. The tears were pouring down on the priceless little bits of smoked salmon, and everyone was being so kind not to notice. This I would treasure always, and they knew it.

Quite naturally, they had all taken a keen interest in the boss's horses. In fact, they had formed their own fan club, roping in their boyfriends, girlfriends, husbands and wives. They put £1 in the kitty every week and when one of the horses was running at a Saturday meeting, they hired a coach, took plenty of eats and drinks with them and made it a good day's outing.

With the help of the rest of the staff the in-tray was looking much healthier, so on the following Tuesday, I decided to take a day's leave, jumped into the Escort and drove down to Surrey to make myself known at the Simon Galloway Stables.

I liked him straight away, he was a nice lad and I could very well see why Ken had insisted on the horse remaining there.

Cleanliness and efficiency were apparent everywhere. The staff seemed to get on well with each other and with the guv. Simon told me that they grew all their own food on the farm. Hay, straw, oats, with plenty of vitamins and minerals, and also molasses. It looked like a thick black treacle to me, but he said it was full of iron and a very good appetiser, linseed, bran and beet pulp.

It was an extraordinary feeling creeping over me as we made our way toward Music Adored's box. I was finding it terribly difficult to realise that she was actually mine, and although of course in a way I felt that I would give the world to have Ken alive and well again, and that Music Adored was still his, it was all so thrilling, so utterly unbelievable that I knew Ken would want me to enjoy every moment and not to be sad in any way, but it was not easy to separate the two emotions, they would entwine, however I tried to keep them apart, but there was one thing that made me feel really good, and that was that Music Adored was at such a good and loving home.

As we strolled on Simon told me that at times he put the horses into the field to be with the sheep and cows, and other farm animals. It sounded almost like a fairy story "they all get on so well together, and enjoy each other's company, makes them feel at home, part of the family."

I could very well see why he was becoming so successful in his career, he obviously cared very much for all the animals and just naturally did his best for them.

The first lot were coming back from the gallops and we watched them come in and get ready to be hosed down. Simon went on "when the ground is hard with frost, he would take the horses down to Camber Sands to gallop along the beaches, and if not too cold have a little paddle, as the sea water was very good for their legs." It must have been quite hard work, loading them, driving down and attending to all their needs, but Simon made it sound more like a B.R. Breakaway Day. "The horses and the lads really enjoy galloping along the sands and although too cold for a swim at that time of the year – they enjoyed having a paddle."

"I don't believe in entering horses in National Hunt races too young, Mrs. Elphick. I think five to six to be a good age to begin racing, then they should be at the peak of their career at about nine or ten years of age."

# Chapter Six

Standing out in a prominent position among all the bills in my Monday morning post, there was a long official looking envelope from Steve's company enclosing legal documents. Steve had put a personal note inside suggesting that he should call the following evening, when he would explain the various papers to me. I phoned him to say that Tuesday was my night at the Judo Club, but I normally arrived home around nine-fifteen and had a late supper, and would be delighted if he would join me then, if it wasn't too late for him.

Before I left home, I put some breast fillets of chicken and a few sliced vegetables in a cook-in sauce in the oven, and speared some jacket potatoes to go with it, and left it to cook slowly.

When I opened the door on my return, the smell that greeted me was gorgeous. It was just as well that Steve arrived almost immediately after me, or there would probably have been very little left for him. This judo lark made one very hungry. At the little bakers, just around the corner from where I worked, they made delicious apple pies, they were so light they just melted in the mouth, and topped with cream, all in all it made a very nice little supper.

The coffee was perking nicely, and to give myself a bit of dutch courage I poured two brandies, and then felt more able to approach Steve again about Ken's accident. He was always so quick to reproach me if I suggested it was anything but an accident.

"Look Steve, I know you think I've got a vivid imagination, but I can't shake off this feeling about Ken. I know he was bothered about something the night before he died, and I don't know if I shall ever be able to prove otherwise, but I am sure it wasn't an accident." Steve was a bit stuffy.

"I think, my dear, that you really must try and put your thoughts in another direction. It is quite understandable, in your grief that you should try and place the blame elsewhere, and I would be the last to be disrespectful to old Ken, but we all make mistakes at times, and it would be much easier for you to accept the fact."

"Had there been the slightest suspicion, I am quite sure the Police would have detected it, and you had the car re-checked yourself." A great help he was going to be. The best thing I could do would be to change the subject.

I had always liked Steve, he had to be nice to be a friend of Ken's and since circumstances had forced us to

## Chapter Six

see much more of each other of late, I was surprised to realise how fond of him I was becoming. What I didn't know at the time was that the feeling was reciprocal and that it was hurting Steve to see me tearing myself to pieces like this.

He went on "…it's too ridiculous Val, there just isn't a shred of evidence that anything was wrong, and one just can't start a private enquiry on instinct, it's too expensive these days anyway. Ken certainly hadn't mentioned anything to me about any problems he was having, but there again, that was understandable, as I only dealt with his personal legal affairs. Any proceedings ascertaining to the business, had to go through Lloyds Legal Department."

Without Ken, the normally smooth running of the office was a bit shattered.

All the staff were doing their best to cope with the additional work but things were beginning to get a bit out of hand, and a rearrangement of the work schedule would have to take place very soon.

I decided to stay late on the Wednesday evening to try and catch up with a few of the more pressing problems, and having got stuck in, the time just simply slid away.

As anyone with a hearing problem would already know, it makes one more sensitive to movement – a bit like a blind person having better hearing, only it works in reverse, so I was aware, without hearing a sound, that someone was in the outer office.

At first, I thought it was probably one of the Security Guards just checking, but then he wouldn't be opening drawers and filing cabinets. No, somebody was looking for something, and the way he was going about it he knew what he was doing.

Very soon he was going to discover that there was nothing worth finding out there, and would in all probability be making his way into the inner office.

When my breathing started to come back on a more even keel, and my heart felt that it was less likely to bust through my ribs, I began to think "God Bless Mick, who had made me sweat my guts out for an hour every Tuesday and Thursday doing press-ups, squat thrusts and break-falls."

I knew I was no Brian Jacks, it had taken me seven years and twenty-one gradings to get my green- belt, but I reckoned that unless that bod in the next room was going to turn out to be a black belt, I may just be able to put up some sort of show.

The only reason I bad gone in for judo in the first place was because I enjoyed my food too much, and it caused a bit of a weight problem, and this was the very first time I had had occasion to be thankful for it.

Goodness, I was so scared I felt I wouldn't be able to stand up, let alone throw anyone, but my brain was doing double-time trying to figure out which would be the best way to tackle him.

Should he come in from the door behind me I could go for an Ippon Seoi-Nage, which would be to grab him around his neck with my two arms from over my shoulders, and to throw him forward over the top of me.

If, on the other hand, he came in from the door at the side, it would have to be a Tomoe-Nage, I would have to grab him from the front, roll backwards on the floor myself taking his with me, throwing him away behind me. Or perhaps a Tai-otoshi would be better, where I could bounce him off the hip.

I was praying that he didn't know his judo.

It's all very well being thrown by a massive hunk of man at the Club, onto a beautifully sprung mat that cost them two-thousand quid, but I wasn't looking forward to coming down on top of one of the filing cabinets.

As it happened, I didn't have any more time to think about it anyway. Fortunately I saw the shadow first, came up with a perfect Tai-otoshi, and my feet hardly touched the ground until I reached Lime Street.

He must have been quite a heavy guy, around fifteen or sixteen stone, but he was a fast mover.

The alarm was raised and a search made immediately, but there was no sign of him. He knew his way around, must have been a professional but what could he possibly hope to steal? No large sums of money or valuables were ever at the office, only files, papers, information. Information, that's what he must have been after. When all the excitement was over, and I began to turn it over in my mind, I wondered if there could possibly be any connection between the snooper, and Ken's accident Could Ken have possibly come up with something, someone had tumbled, and he had been put out of action. Quite possible, but what?

This was it, there had to be a connection, but the one thing that was certain at this stage, was that the guy in the office was quite definitely not the person talking to Chopper at Stratford the night of the play. Their builds were the dead opposite.

I phoned Steve "Not just to say I told you so, but because I really need a friend".

"You have got yourself one, what's the next move?"

"What about an East End pub crawl, does that take your fancy?"

"Marvellous, I'll pick you up in about thirty minutes."

"Best to start at the middle, the pub nearest the place where the two men met, and fan out from there."

So the first call was The Sparrows.

"What are you drinking Val?"

"Would it be better Steve, as I know the area so well, if I stayed with tonic waters, and drove, we could save a lot of time taking the back doubles."

We wound our way through an assortment of Kings – Alfred, George and Harold. Queens, – Victoria, Arms and Heads and Princes Albert, George and William, with a few Lords, Henniker, Clyde and Palmerston thrown in. Very patriotic the Cockneys. They also seem to be very fond of Green – Men, Gates and Dragons. Finally, as it seemed appropriate, The Chevvy Chase.

Fifteen tonics, ten pennies spent, and not a murmur or sign of anyone who looked remotely like Chopper's friend.

This was not altogether surprising, as all the pubs had been glamourised in the past few years, and that meant subdued lighting, notably red, orange or green, and it was almost impossible to recognise the person you walked in with, let alone someone you didn't know, and was looking for

Perhaps unfairly, I had always thought of Steve as

being a bit of a stuffed-shirt. Had I known that he and Ken were ex-Barnardo boys, I may have understood more readily the struggle it must have been for them both to have reached the high standards in their respective chosen professions. But tonight was different. Steve was beginning to let his hair down a bit more, becoming more human in a way, and making a few mistakes, which was unusual.

It was inevitable, as a practising lawyer, that Steve must have had to have been involved with investigations from time to time, but he would always employ Solicitors Agents for the job, and he himself had yet to learn the most elementary rudiments of investigation. Rule one, order a gin and tonic, fill it to the brim with the tonic water, take one sip, and either conveniently knock it over, or give the nearest rubber plant a treat.

The easiest way to detect where those on the Wanted List hang out is to look at the plants, they are right bushed best part of the time. I made a fairly neat job of parking Steve's car in his garage, and cleverly disengaged myself from his goodnight kiss, and was lucky enough to catch a cabbies eye to take me home.

# Chapter Seven

Saturday morning, and I was looking forward to a day's racing at Newbury.

The teasmade came up with the goods right on time, as Richard the newsboy was popping the Racing Post through the letter-box, and he rang the bell to set my flashing lights going, so that I should know that it had arrived.

This was sheer luxury, to sit up in bed, with tea, to absorb and compare the knowledge of the experts. I had so often said, I would rather go without lunch, then give up my Racing Post, or in the old days The Sporting Life, and this was the one day of the week when I could take my time and digest the various opinions, together with those of the Scout of the Express, and Captain Heath of the Mail. Not that I would necessarily back the tips they gave, unless of course it was one that could not be opposed. Better to back a favourite that wins, than an outsider that loses the old hands at the game used to say, but I very rarely made a choice until I had seen the horses in the parade ring and watched them go down to the start. A jockey can easily lose a race by the way he takes the horse down to the start, it is very important to watch that.

It is necessary to know about form of course, but I try not to get too involved, and prefer to rely on my own instincts, and to know how the horses looked and behaved in the paddock, and it had not let me down too badly in the past, although I must confess there had been many times I had lent over the rail of the parade ring, and tried to lip-read snippets of conversation between trainers, owners and jockeys, which sometimes had helped me reach a final decision. There must have been something in it, because I always noticed that when I watched racing on television, that it was better to just have fun bets. I could never seem to find the winners as I did at the Course.

My friends pulled my leg saying that I did better than they with racing on television, as I was unable to hear what the Tipsters were saying, and that had to be in my favour.

Tipsters have always come in for a lot of criticism, rightly so in some cases, but the game would come to an end if they knew every winner. In the old days, when tipsters sent their 'certainties' by post, the system was to split the country into however many horses were running in the race. If there were eight runners, then a map of the British Isles was split into eight sections, with a different horse tipped for each section, so someone, somewhere, had to win, and they lived in hopes that the population as a whole, each received their fair share of winners. It didn't work out that way of course, and then the letters started to pour in: "…you couldn't tip your grandmother out of bed…" etc. I well remember as a youngster the laughter and fun we had when the great Prince Monoloulo, such a colourful figure on the Racecourse, would often come home with my father after a day's racing, for a meal, or a drink and they recalled either the praises and adulation that were rained on him if he found a winner, or the various suggestions made as to how he could dispose of his tips, if they failed.

A much more sophisticated method prevails today, but it was fun then.

It was very often possible then to lose money without having a bet. Pickpockets on the course and three card tricksters on the trains. Very often when a new face appeared on the trains they were broke before they reached the course, or if they managed to escape that hazard, they would more than likely pull a wallet out of the back pocket, sometimes bulging with notes, to place a bet on the first race, but it would be gone by the second.

This was the weekday occupation of the boys who did Petticoat Lane on a Sunday. It was always said that, together with a wallet, they would nick a gentleman's handkerchief out of his top pocket at one end of the Lane and sell it back to him at the other end.

All kids stuff compared to the high technology of the crooks of today but none-the-less it made them a tidy penny or two, in some cases invested in property bought for a few hundred pounds, and today the land alone worth several thousands, enabling the three card tricksters, and bookies-runners, of the pre-betting office days, to live in comfortable retirement, very often spending best part of the winter in the Bahamas soaking up the sun.

They can sit back in their Parker-Knoll and muse over the battle of wits they used to have with the local coppers who tried to book them for taking bets.

One East End copper, who was well known to them, was very aggrieved that the bookie's runners always managed to outrun him. He was determined to book them, so one day he borrowed a barrow, dressed up in his old togs, rolled up his sleeves and wore a choker and cap, and paraded the streets as a rag and bone man, thinking his disguise was complete, but the boys were wide'o he hadn't fooled them one bit, and as he made his rounds he was piled high with all the old rubbish they could lay their hands on. old tables, chairs, rags, any old sinks, bins, with a few rotten tomatoes thrown in for good measure.

I had a leisurely bath, instead of the usual quick shower, put on my comfortable trousers, suede jacket and 'racing shoes', swung my binoculars over my shoulder, and was ready for the "off". It did occur to me that Steve may not be feeling his usual self, and I thought it better I should give him a ring before I left. I really had enjoyed my night out with him, although it had not brought any immediate results.

Bruur... bruur... bruur... no answer, I turned my amplifier on full blast, perhaps I was not hearing him.

"Yes."

Chapter Seven

"Steve, how are you this morning?" he sounded terrible

"Oh Val... Leave me... for about a fortnight"

"What's up Steve?"

"My head feels that there are about ten little men with hammers all in competition as to who can bang the loudest."

"Oh you poor dear, shall I come round and fix you something, coffee and an aspirin."

"No thanks Val, I'll manage."

I guessed that the last thing Steve would want would be for me to see him like that, he always looked so well turned-out.

"Oh my God, how do those agents, who put so many drinks down on their expense accounts manage, they couldn't possibly drink them all".

"Any more of your bright ideas Val, and with a little more practice, One Solicitors Agent will be minus a client."

"In future I think I might find time to do a little investigating of my own, I might even enjoy it. There must be quite a bit of satisfaction turning something up. I will have to start with the easy ones though and work my way up. You've dropped me right in at the deep-end, and if this Chopper bloke has anything to do with it, he knows how to handle himself."

The traffic was light, being Saturday, and I was making good time to Newbury.

I found the winner of the first.

Simon had finished saddling up the only horse he

had running at Newbury that afternoon, and noticed me as he was leading the horse into the parade ring.

"Would you join me for a drink after the fourth, Mrs. Elphick, I would like to have a word with you before you go?"

"Yes I would like to, shall I meet you in the bar?"

"And by the way, good luck."

My heart felt that it had dropped to the bottom of my tatty old shoes. My immediate thoughts were could there be something the matter with Music Adored. I was silently sending up a little prayer – "please let her be alright."

Simon's horse won comfortably, and Simon notched up another winner for the season.

It must have been one of the longest hours ever, normally there was hardly time to get everything in between races. Putting on bets, drawing winnings, catching a quick cup of tea, and what I enjoyed most of all watching the horses in the parade ring, trying to pick up little snippets of information.

By the time Simon arrived I had already ordered a coffee for myself, but he insisted I should have a brandy in it as the afternoon was turning a bit nippy. In my state of mind, this made me feel even worse, I felt sure he was trying to soften some sort of blow, and I could only think of Music Adored, so it took me quite by surprise as tipping the brandy in my coffee he said "I had a young fella down at the Yard the other day, said he was a reporter, asking a lot of questions about you."

I could not believe it "Asking questions about me, how could anyone possibly want to know anything

## Chapter Seven

about me?"

People were continuously interrupting to shake Simon's hand obviously the ones who had backed his horse, and he eventually went on "I thought it rather strange at the time. I have never seen this chap before, certainly not at any of the Meetings, and they usually make themselves known to us, particularly when they want some information for a write-up in the next days paper, and normally it's the horses the questions are all about, but this one kept putting questions to me about you."

Naturally I told him I do not discuss Owners with Reporters, and it also struck me as being rather odd, particularly as he said he was from your local paper, and seemed to be completely uninterested in Music Adored. I would have thought it would have been better for him to have contacted you locally.

"I thought I had better let you know, put you on your guard if he should call to see you."

"Thanks Simon, I'm glad you told me."

"How is Music Adored?"

"Fine. When the ground's right we will enter her for another race. I will be in touch." "By the way, I should give the local a ring if I were you, see if they did send somebody down."

First thing Monday morning I phoned the Gazette – a young girl answered, I could visualise her at about nineteen years of age, with a miniskirt that looked just like a pelmet, dyed hair, loads of mascara and very long fingernails with a nasal voice "Tracy here, can I help you?"

"Could I speak to the Editor please?"

She was horrified – the Editor – the big boss – certainly not. "I will put you through to his Secretary" who then transferred me to a reporter, who then transferred me to another reporter who then had me put through to the sports section, and finally back to the Secretary. I was getting uptight – please let me have a quick word with the Editor – and much to my surprise it worked.

"Look, I am sorry miss, but no-one here has ever heard of a horse called Music Adored let alone the owner. We certainly did not send a reporter to the stables, and besides it would not be our policy to ask personal details of anyone but the person concerned. We have our reputation to think of you know."

"Of course, thank you for speaking to me."

I had a busy morning ahead, and for the time being, at least, I had to concentrate on what I was doing, and put this strange incident out of my mind. Nevertheless, during the day it kept floating back from time to time.

I contacted Steve.

"Look Val, I think perhaps it would be as well, if I had a word with the Agent I normally call upon, this is more in his line."

"No please don't do that Steve, it will only involve a lot of expense, and there's so little information we can supply him with at this stage." I didn't tell Steve, that I felt within my heart that I must personally be involved as a sort of thank you to Ken. No doubt the initiative that was spurring me on was really revenge. Had someone been directly responsible for Ken's death, then

Chapter Seven

I wanted to be party to seeing them pay the penalty. I may be an amateur, but I certainly had the incentive.

I decided that the next day I would start cross checking all the details of the horses underwritten to the firm and try and discover if a fraudulent claim had been made.

I checked, re-checked, double-checked and treble-checked.

I had all the Veterinary Certificates verified, and I paid special attention to all the markings, and went over and over again the details of the two Australian horses with similar markings, one cost $125,000 and the other $35,000. A claim had been made for the first horse and had been paid. It was later discovered that a horse entered in the name of the second horse for a race at Melbourne, was in fact the first horse who was supposed to have died. The case was currently with the Legal Department, and I gave very particular attention to this one, but could find no connection at all with Chopper and his associates.

Surely there must be something here that I am missing if it was important enough for an intruder to want it. I would just have to press on, it was probably staring me in the face and I wasn't seeing it.

# Chapter Eight

Saturday, and I got my usual flash of lights from the paper boy and rushed down to pick up the Racing Post. There were a few letters on the mat, I could see the telephone bill there, but the top envelope bore the stamp of the Evening Standard, and I knew, before I even opened it, that I was in with a chance. My fingers trembled like mad, and I was having one hell of a job getting the blooming thing open, and then, that first beautiful, wonderful word "Congratulations" and I knew, yes, I had won the Evening Standard Competition sponsored by United Racecourses Ltd. for a champagne lunch, a grandstand seat to see the Sandown Salver, the world's top jockeys competing in the Great Britain v. U.S.A. jockeys "rubber" match at Sandown Park, together with fourteen months free Membership at the Course. I felt I was going to burst with joy. I knew I could do it.

Although I had always admitted National Hunt

Racing was my favourite, I had always taken a keen interest in the flat, and knew very well that I had answered the questions correctly, but it was the slogan that had me worried, which was about Sandown Park, so I just put what I believed to be true, and that was that it was well-designed, attractive, sporty and very efficient, and the judges must have been satisfied.

So instead of the usual Saturday morning stint of spreading all the papers around, picking out winners (or losers) I decided for this very important day I really should invest in a new pair of 'racing shoes'. it, has always amazed me how the ladies managed Royal Ascot in their shoes. I always liked to put on my comfortable togs and feel free and easy to enjoy the racing, but for Wednesday my feet were going to have to suffer. As it happened it was all so exciting I hardly noticed that I was tottering around on reasonably high heels. Quite legitimately this time, someone at the local Gazette with a keen eye had seen my name in the Evening Standard when the winners were announced, and had turned up to take a picture of me sticking my "Members Car Park" ticket which Sandown has been thoughtful enough to enclose, on the windscreen of my car, and then the photographer together with a few of the neighbours, came out to wave me on my way, all wishing me a lucky and happy day. They knew I had had a bit of a rough time of late and they seemed genuinely pleased to see me looking so happy.

Arriving at the course around 11 o'clock I was told to make my way to the Secretary's Office, and was then escorted over to Members and taken up to the Banqueting Suite. I was introduced to Falcon of The

## Chapter Eight

Standard who began introducing me to all the people around. I obviously knew some of the faces, but I also knew I was going to make a complete fool of myself, because the lighting in the bar was not suitable for lip-reading and it is only possible to lip-read one-to-one. It was time for me to confess. I took The Falcon aside and said to him, I know I should have mentioned this to you before, but I don't hear.

"You what?"

"I don't hear."

He turned to the Racing Manager and said "we've gone to all this trouble and she's not going to hear a bloody thing"

"I do lip-read."

"What was the greatest horse ever?"

"Arkle in my book."

"You'll do."

I asked him "please just don't expect me to be able to communicate with all the people at the table – I can manage one-to-one".

At noon the first champagne cork popped, and it was flowing all afternoon, as freely as the never ending racing tales that racegoers always enjoy telling and listening to, a most important part of the pleasure.

The atmosphere was electric. Everybody, who was anybody in racing circles was lunching in Members that day, and even the most experienced of racing people were becoming very excited about this very special day.

Sandown was looking it's best, always lovely, it seemed to have a special sparkle, and after twenty days of rain, the wettest month so far, I was not the only one on my knees praying for a nice day. The organisers needed a gate of at least 7,000 to break even. The jockeys arrived, all in their team jackets. The British Team, sporting the Union Jack, and the American the Stars and Stripes. I.T.V. pulled out all the stops, and sent a complete team, John Oaksley, Brough Scott and Derek Thompson were all there, as there were so many people to be interviewed.

The splendid lunch was over, and we were all preparing to get down to the main business of the day – the racing. Ah, this was the life, could anything be better than this? Good company, good food, and now leaning on the rails of the paddock anticipating the best competitive racing.

Finding the winner of the first two races was not easy, but who cared, there was plenty of time to put things right by the end of the day. The two-forty-five was one mile six furlongs, and the start was in front of the Grandstand. I loved being out there, jostling among the bookmakers, comparing the prices, and feeling great if I could get in at a price one point above what most of the bookies were laying. I thought I would walk over to see the horses go into the stalls, and suddenly, there he was, right in front of me. Chopper.

Well I'll be blowed, all that time scanning the East End pubs, and here he was, on a plate, so to speak. But what could I do now that I had found him? I could hardly confront him without the slightest shred of evidence that he was mixed up in something, I know not what, just my feminine instinct doing overtime.

## Chapter Eight

Pushing through the crowds making their way up to the stands to watch the race, I finally made the lift up to the second floor. If I hadn't already lost him, this would be a good vantage point to keep an eye on him, see if he made any contacts. I had a good pair of glasses, I needed them as I was unable to hear the commentary, and this was in my favour. It was possible to have an all-round view from the Members Dining Hall. One side of the room covered the Course, and the other the Paddock and Unsaddling Enclosure.

I kept my eyes glued to him best part of the afternoon, and then I was taken over to have my picture taken with Lester Piggott and Willie Carson. Lester led the English Team, and Willie scored the most points that day. The final score was Britain 40, U.S.A. 26.

It was at this time that I was given the opportunity of asking Derek Thompson if he would allow me to nominate him as "Golden Lips of the Year" for Sports Fans. He is so good to lip-read and has such good expressions. He won of course, because I got all my mates to vote for him.

I received a few cards saying congratulations when the photo was in the next day's Standard, which were all very nice, but the one I liked best was a remark from a neighbour, who had worked hard all his life and done well for himself but still managed to keep his Bethnal Green accent.

"Here Val, I sent my little typist out for a Standard yesterday afternoon and cor-blimey – the others in the office gathered around and seeing the address asked "Do you know her?"

"Know her, I thought I was the love of her life and here she is having it off with Lester Piggott *and* Willie Carson."

# Chapter Nine

Simon contacted me on the Tuesday, to say that Much Adored needed a race and he had entered her for a two-and-half mile chase at Lingfield the following Saturday. The boys and girls booked their coach, and some of my enthusiasm rubbed off on Steve, who was not a racing man, but decided to come along to see what it was all about. Fortunately the rain held off and racing was on at Lingfield on the Saturday, a course that very often becomes waterlogged with heavy rain. Steve wanted to take his car and stop for lunch. I would have liked to have gone in the coach, but he had been so good, I thought it best to fall in with his wishes, so I phoned Jim at the Blacksmith's Head at Newchapel, about a mile down the road from the Racecourse, a favourite with race goers as the food is always good and exciting and everyone receives a warm welcome from Jim and Joan, and as usual, their timing was perfect. Jim had seen to it that lunch was prepared early to enable everyone to get to the course in plenty of time for the first race.

The scene in the parade ring was truly colourful. The springy green turf, the lovely trees and flowers, and the ladies all togged up for the day in case they were caught by the television cameras, this all blending beautifully with the jockeys in their bright and jazzy colours. Amanda Neale-adamson was giving her jockey a leg-up on Ongar, one of her father's horses, which was being ridden that day by Lord Hatherley-Jones, who normally only rode his own horses, but agreed to ride Ongar for Amanda when she asked him, as the jockey booked had taken a nasty fall in the previous race, and the Doctor would not pass him fit to ride. His Lordship was liked by the racing public and bookies alike for his courage, good nature, and just for being a good sport. He must have broken every bone in his body at some time or another, but he kept going, and even rode in a couple of Nationals. He was very tall, thin and had a pale complexion, and as Saint Paul's colours were predominantly white, it emphasised the paleness even more, and some joker in the crowd shouted "blimey guv – you look like a bleeding bottle of milk'

Music Adored ran a good race, which she needed, and came in third. Ongar was coming up to the post, last of the seven, and Hatherley-Jones was getting a terrific cheer from the crowds, just for managing not to part company, which he acknowledged with his usual beaming smile, but it was becoming increasingly obvious with every stride, that Ongar was not going to make it to the post. His jockey pulled him up, and was about to dismount, when Ongar just bowled right over, taking Hatherley-Jones with him, who gashed his cheek badly as he hit the rails. The horse was still down, but

# Chapter Nine

the jockey got to his feet and propped himself up against the rails of the Grand Stand. The Vet ordered the screens to be brought, and depression settled temporarily over the crowd. It is always a sad and depressing atmosphere when a horse has to be put down. I had known the Hon. Amanda Neale-Adamson, Owner, Trainer, Breeder, slightly, and more intimately when Ken had his horses at her Yard.

She had married Stuart Neale-Adamson, a very ambitious businessman, when she was nineteen. They appeared to be totally unsuitable to each other. Amanda's heart was obviously with her horses, she was a good trainer, and knew and did her job well. It was generally thought, that Stuart had married her because her father Lord Paul Franklinford, a nutty do-gooder, had all the right connections. Commonly known as Saint Paul, the more murders someone committed, the harder he fought to get them released from jail, and I had heard around racing circles that he was glad when Stuart proposed, as with Amanda's lack of femininity he was unable to visualize the young men lining up for the privilege. They lived on a lovely estate near Newmarket. The Canadians had taken it over for their Officers use during the second world war, and restored it to its original beauty when they left, adding all the latest mod cons. When I went there with Ken once, I was very impressed by the layout. There was a long circular drive, up to the house, boarded by trees and flower beds, which surrounded the lawns they sometimes used as bowling greens, there were sunken gardens, tennis courts, swimming pool and lake, but this was all before we reached the really impressive part, the Stables. They were really something. On sight, it looked as though there was everything there an owner could possibly wish for, but something must have been lacking for Ken not to have wanted to have kept his horses there. I walked over to offer my condolences to Amanda. She cut me dead. I knew how she was feeling, particularly as this was the second horse to die on her within a few weeks. I had been getting the cold shoulder from her since Ken had taken his horses away from her Yard. I had no idea why Ken had made, what seemed to me, a sudden decision to put his horses with Simon Galloway, but knowing Ken it had to be a good reason, and certainly a decision he must have given much thought to. He was not impulsive by nature. Poor Amanda, it was impossible for her to disguise her feelings, she aged visually within minutes. I saw her pacing up and down outside the changing rooms, waiting for Hatherley Jones to reappear. He had had to see the Doctor, to give him the once over, and put a dressing on his face, and he had then gone in to change into his outdoor clothes.

He was hardly out of the door when Amanda pounced on him, but I doubted if there was anything he could tell her to put her out of her misery. Only the Vet would be able to come up with the answer to that one.

Ongar must have been a disappointment to her all the way along the line. It was unlikely that he would have ever hit the headlines, if he had not dropped down dead in front of Tats.

One day, perhaps a good jockey may have been able to have got him first past the post, but in all probability the best part of the field would have had to have fallen for him to accomplish it. Perhaps it would have been better, if I hadn't approached her, it must have been doubly agonising for her to see Music Adored run such a good race, particularly as she had bred her.

She obviously didn't want to lose any of Ken's horses to Simon, but it must have been extremely painful to see Music Adored go, particularly as everyone seemed to be of the opinion that she was going to turn out to be an exceptionally good horse.

In the meantime, Saint Paul was absolutely wallowing in all the sympathy he was getting from his old cronies, patting him on the shoulder and muttering a few inaudible words. Amanda looked livid, it was she who loved the horses. More than likely the old man would not have known which of his horses had died if it's name hadn't been in the race card and over its box at the Yard.

"Well, Val, this is it, your first time in the unsaddling enclosure as an owner, how's it feel?" "Wonderful" "Let's hope it's the first of many." And what a thrill it was. True it was only third place, but Simon was more than pleased with Music Adored's performance, and her precious blanket was being given its first airing. The boys and girls bad been warned not to put their money on to win, and at the same time Simon happened to mention that, a horse jointly owned with his brother, and being ridden by him, was fit, and barring accidents, was in with a chance. It obliged, and the gang showed a profit on the day. They were as excited as I was, and I would have loved to have shared their joy on the coach home. There is nothing like the relaxing time of strolling back to the coach together, exchanging all the little tit-bits of the afternoon, feeling tired, but a nice tiredness, especially if one was coming out on the right side. Then the fun of unpacking all the eats and drinks that always tasted so much nicer picnic fashion.

Philip Hunt, Saint Paul's Secretary, and a first euphonist in the Band, was making his way across the "Owner and Trainers" car park when I spotted him.

I managed to catch up with him.

"Philip, please tell Lord Franklinford how sorry I am about Ongar." "Oh and by the way, I was hoping I would see you today. Earlier this week I was turning out a cupboard at the office where Ken kept his personal papers, and I found a file with letters and cuttings about the Band. I think the Band Secretary should have these now, so if I could drop them into you, would you hand them over Band practice night for me please?"

He seemed miles away, I was not sure if he was taking in what I was saying, but something clicked and he said "Yes, of course. In actual fact, I shall be passing your house tonight at about eight-thirty to nine, if you like I can pick them up then."

"Fine, I'll see you then."

Steve had enjoyed his day, he was getting the fever. Said he would like to go again.

"Val, do you mind if we made it a quick meal on the way home, I have a lot of work to catch up with?"

"If you would rather Steve, I can make myself something when I get home, and I am rather tired after all the excitement."

## Chapter Nine

He looked as though he was quite pleased with my suggestion.

"I'll tell you what, I will give you a ring early in the week, and we will make up for it with a slap-up dinner one evening, how's that?"

"Great, I will look forward to it."

He just couldn't wait to spend his winnings. Beginners luck, he will learn after a few more trips to the races.

After I made myself a snack, I tried to settle down to read, but I was too excited. Excited and puzzled. Puzzled that two of Amanda's horses should die quite suddenly. I turned it over in my mind. Could there be any connection. Of course, it is possible for two horses to die in similar circumstances from the same Stable, but most unusual.

I had always known that Ken could shut up like a clam if he wanted to, and when I tried to pump him about taking his horses away from Amanda and sending them to Simon, he just said he was an up and coming young man who would do well in his career, but I was strongly suspicious that there was very much more to it than that. Ken was not the type to rush at things and must have considered the move very seriously.

Philip pressed the door-bell at eight-forty-five, dimming the lights. The flashing light unit has the reverse effect at night, instead of flashing, the lights dim.

I put down my book, puffed up the cushions and made may towards opening the front door.

I found it almost impossible to believe that a person's

appearance could change so dramatically, in such a short space of time. He looked absolutely washed out and haggard.

"Have you got time for a coffee Phil?"

"Thanks Val, I could do with one."

I had noticed that he had not been his usual happy-go-lucky self for some time, and I had suspected that he was having an affair with Amanda, and just thought that as her father employed him, that perhaps he was concerned about his job.

"How about a drop of brandy in that coffee, I'm having one?"

"That would be lovely. It's been a tiring day."

I let him take his time, and allowed the warmth of the room, the coffee and brandy to relax him a little.

"Any news about Ongar?"

"Not yet, there will have to be a post mortem. Amanda's feeling pretty cut up about it."

"I can imagine."

"One way or another, it's been a pretty lousy day for me too."

He held his head in his hands, and I could see he was near breaking point.

"Is there anything, I can do. Sometimes it helps to share a worry?" His hands were shaking...

"I've been such a bloody fool, and now Amanda's putting the pressure on. I just don't know which way to turn. Whichever way I go I'm going to be in trouble it seems."

I was amazed at this outburst, it was so unlike Philip.

"Philip, have you stopped to eat anything today?"

"I had a sandwich at lunch-time."

"Let me make you something, what about a spanish omelette and some french bread?"

"Sounds lovely, but I don't want to put you to any trouble Val."

"Won't take a minute or so."

My conscience was beginning to prick. My idea about offering him food was to get him into a mellow mood and talking, and maybe I would learn something about the horses in Amanda's care. I soon put the omelette together, and left him to enjoy it, while I went to make some fresh coffee.

He looked a little more relaxed when I returned, and I was kicking ideas around in my head as to how I could pump him without it looking too obvious, and finally decided it was better to break the ice with something a bit nearer home ground…

"How did you come to join the band Philip?"

"It goes back a long way now, but when I was a kid of about eight, Mum and Dad talked me into joining the Salvation Army Band, as they wanted to get me out of the way on Sunday mornings." He went on "I started on cornet, and worked my way up to flugel horn, then baritone, and finished upon on the euph."

"You know, I can remember when I was about fourteen, saying to the Bandleader "I'm a man now I'm not going to play with kids any more" and I walked out, "but I loved the brass band world too much, and after a few weeks of not playing, I applied to the local band for a position."

I could see the tension was beginning to lessen, so I stayed with the same theme "What about your first solo?"

For the first time that evening, a smile gradually broke across his face. "A slow tuneful melody – Watching the Wheat. It was a good job they were watching the wheat and not my legs, they were like jelly."

"It's a different story today you must have nerves of steel, double tonguing, triple tonguing and control of the four valves."

"Hardly nerves of steel, in fact, today I feel a complete wreck."

I hadn't meant to say the wrong thing, I must have trod on the wrong nerve, his hands went up to cover his face, he was swaying to and fro… "Oh Valerie, I've made such a mess of my life, what am I going to do?"

I was getting a bit hot and bothered, I hadn't anticipated this after the coffee and omelette "You've had a tiring and upsetting day that's all Philip, things will look different in the morning."

"They won't you know. I had enough on my plate before, but now Amanda's putting the pressure on" I let him go on "You've probably already guessed about Amanda and me. I didn't really feel too badly about it as far as Stuart was concerned. They were never suited to each other. He only married Amanda because he thought that Saint Paul knew all the right people, and it would help him climb to the top, which no doubt it did, and he's away best part of the time. As you know you

## Chapter Nine

can always find a pretty girl if you have the money, and he has. No doubt variety is the spice of life as far as he's concerned, but what I didn't reckon on, was becoming as fond of Amanda as I have. I know she looks a bit of a plain jane, and when she gave me the big come-on when Saint Paul first introduced us, I thought it a bit of a joke, but once you get to know her she really is a very nice person, and somehow, I don't know why, but we just clicked. Now I feel I don't want to let her down."

I just did not know what to say – "I'm sure you won't."

I poured more coffee, without brandy this time as he was driving, he seemed lost for words.

"Unfortunately, it's not as easy as that. Look, if it's not boring you too much, can I tell you something about it, I feel I've got such a lot bottled up."

Boring me? I was all agog…

"Go ahead, Phil, if you think it will do any good, you know I would be discreet and not repeat anything told me in confidence."

"I can't really tell you all that is worrying me, but on top of everything else, one day last week Stuart must have decided not to go to town early as he normally does, Amanda thought he had already left but he must have been in his study making some telephone calls. As you know he has done very well for himself, and I have to give him his due, he has worked hard for it, but he's got this thing about getting a knighthood, it has almost become an obsession with him, and Stuart's no fool, he knows, better than most, that the best way to go about it is to become involved in all the big charity stuff. So for a kick-off he has offered the use of the house and grounds for a fashion show, to be given by one of the big Houses, and to be televised. Well, he suddenly got it into his head to phone over to the stables to remind Amanda to get herself a dress for the special night.

"Apparently, it came out afterwards that he had been ringing for some time without getting a reply, you see the lads had taken the horses out for exercise, and Stuart thought that the stables must have been left without anyone in charge, and although he very rarely makes an appearance there, he would be only too glad to catch someone out if this very strict rule was broken. Too much expensive gear about to be left unattended?

"…Well, to cut a long story short, and to put it in a crude sort of way, he tumbled across Amanda and me having a romp in the hay so to speak, and to put it mildly he was not very pleased. In fact he was extremely angry, calling Amanda a silly, stupid cow. He didn't seem to be personally offended, it appeared to be more worried that it could have caused a scandal and that it would have made him lose face with the business associates that he said he had fought so hard to cultivate over twenty years.

"Honestly, Val, I thought for a moment he was going to kill her, but I suppose he hasn't got to where he is today without learning how to control himself in adverse situations, and he finally calmed down. "But he couldn't just let it stop there, he went on and on really insulting Amanda, saying how ugly she was, and that he was going to make sure to hide her behind a pillar on the night of the fashion show as it would never do for a roving camera to pick her up, and then added, if the

## Chapter Nine

smell was anything like it normally was when she returned from the stables, they would rove the other way. He could not have been more insulting. There was no need for that, he has not been a saint himself."

I refilled his cup with coffee…"How about another small brandy?"

I could run him home if necessary, he didn't live far away.

I had guessed that this alone had not reduced him to the state he was in.

"It is good, and I do feel better for getting some of it off my chest, and having started I may as well tell you what has happened today, if you will allow me…"

Allow him, I couldn't wait.

"Well, I know Amanda was very upset about Ongar, and I am trying to make allowances for that, but she has now come up with some scatter-brained idea about leaving Stuart, and starting up elsewhere, and she wants me to go with her. You can just imagine how Saint Paul is going to take that. I'll be for the chop that's for sure."

He hesitated… "You see… I know I should not really be saying this to anyone, but we have known each other for a long time, and I know you would not repeat it, but Amanda seems to think that Stuart is having his revenge by killing her horses, and trying to ruin her reputation.

"I have told her that I am sure it's not true. and I really do believe he feels quite proud when Amanda gets good write ups in the press, but you see one or two strange things have happened at her yard recently, and I know its nothing to do with Stuart, but I can't convince her of that."

It shook me to the core, I could not believe that I was going to hear something without having to probe, which I am not very good at, and I was all ears – switching my two hearing aids to the highest level.

"It all started a few months ago, first Daffodil III died, as you know, then the office at the yard was turned over, and the tack room, all within a week.

"You remember Mr. Stimpson, Saint Paul's friend, he then wanted his horse Ongar II to race at Auteuil. It had been entered, against Amanda's wishes, she did not think it was ready for the race, but a week before it was due to go it punctured its foot, and the blacksmith had to put on a protective pad.

"You see, when a horse punctures or injures its foot the blacksmith has to clean it out, apply stockholme tar, pack it out with cotton wool and put on a protective leather pad before he can put the shoe on. So Amanda strongly advised Mr. Stimpson that they should withdraw, but he is the owner and has the last word and he was not having any, and insisted that the horse should be sent to France.

"Tom the head travelling lad was keen to go, as he has a little filly of his own over there, and Saint Paul said he would go along for the ride, so Amanda said that as shoeing was a very expensive business, it would be as well to race in heavy plates, as the light aluminium plates fitted for racing were unsuitable for the road work which is essential, as you know, for horses to get used to people and noise, and so the heavy plates have to be put on again when the race is over.

"I have known her to race in heavy plates before, a lot

of Trainers do if they enter a horse just to gain experience, or if it's in need of a race, but old Stimpson would not allow it, he said it would ruin the horse's chance and that the light plates would have to be put on for the race. Anyway the most extraordinary thing happened on the following Monday when they had all returned from France Ongar II shed his plate, the one with the leather pad, and after saying a few words about French blacksmiths, Amanda phoned her own farrier to put another plate on. She had a look at the injured foot, and decided that it was no longer necessary to have the leather pad put back and slipped it in the pocket of the grubby gilet she lives in.

"It was a few days after that one morning she found that someone had been turning the office over. I know she doesn't always look a fashion plate herself, but she does keep her office and stables in meticulous order, and she noticed that a couple of the drawers were not quite shut, and that some of the papers were just a fraction out of place. She doesn't keep anything too important there, puts most of it in a strong box at the bank, in case of fire, so she was not particularly worried, just had the locks changed.

"Later in the day she was told by Tom that the tack room had been turned upside down, which seemed ridiculous. However, a few days later she happened to put her hand in her pocket for a tissue, and came across the leather pad. She pulled out the cotton wool, and subconsciously realised that it was a bit heavy for cotton wool, and tore it to pieces, really just to see what French blacksmiths put in their padding, and believe it or not there was a very neat little package containing a white powder – probably worth a pretty penny street value.

"When you come to sum it up, horses are the only animals allowed free passage in and out of the country, no quarantine for rabies or foot and mouth if all is well, and Ongar II was a big nine-year old with a very deep well to his foot to house the extra goods.

"I think the best thing she could have done, would have been to have come out in the open about it, but no way can I persuade Amanda to do so, she insists on keeping it and saying nothing about it She maintains that she could have very easily thrown the padding away without knowing what was inside, it was only sheer luck that she was so inquisitive. Well, that's her story and she's sticking to it." "I have tried to warn her how dangerous it was becoming. She was staking everything but would not listen to sound advice." "I told her, it's just not worth it Amanda, it could ruin your whole future, and you know as well as I do that she is a good Trainer."

I was completely staggered… this was a turn-up for the book as far as I was concerned. It was the last thing I had suspected, but I had to agree with Philip about Stuart.

"I really cannot see what Stuart would gain by damaging Amanda, it doesn't make sense to me."

He was quite animated. "Of course not, it's a ludicrous idea."

He got up from his chair and I noticed he was much more relaxed and looking a little like his old self.

"Well Val, I must be on my way."

He was dithering… "Look Val, I feel I have been a bit

unfaithful to Amanda, spilling the beans like this, but it's really because I am so worried about her. You wouldn't mention anything would you?"

"Of course not, you know me better than that Philip. We will forget you even called here tonight o.k."

"Thanks for the coffee and the omelette, I feel much better now" but he didn't look it. It would have taken more than what he had told me to reduce him to the nervous wreck he had made of himself. I wondered if he knew more about the package than he was letting on.

"Would you like me to run you home?" – I was a bit worried about the brandies but I think he would have been under the limit if he had had no previous drinks.

"No thanks, I am o.k."

And after a most extraordinary night, that I could never have anticipated, I said "Goodnight Philip, take care."

# Chapter Ten

I had a bit to come back from three coming up out of a Yankee I had put on a couple of day previously. They were all favourites, so there was no mad rush to the bookie's office, but as I had managed to catch an earlier train than usual home from work I thought it a good opportunity to collect, and get a price about a horse I fancied for the King George on Boxing Day.

George, a horn player in the band was behind the counter, looking like every favourite had turned up at all three Meetings.

"What's up George, it's full of gloom and doom in here, have the bookies been taking a beating today?"

"Worse than that Valerie, there has been a knockout. Came up at four-to-one, Johnny and the lads are down at the Duke's Head drowning their sorrows."

"Do you think he would mind if I butted in, I wanted to ask him about an ante-post bet."

"Course not, he'd be pleased to see you."

Johnny is very tall, about 6'6", a mass of black curly hair and very very handsome. In the summer he wears silk-monogrammed shirts, unbuttoned almost to the waist, and sported three gold chains around his neck, one bearing a gold race horse, and always looked to me as though he would be more at home doing a James Bond scene on a film set, rather than shouting the odds on a bookmakers stand.

I have seen ladies at classy meetings such as Royal Ascot, stagger up to his stand in their high heels, flutter their false eyelashes, and have a pony each way on a old rag (outsider). Just what Johnny liked, and he knew exactly how to play up to them. He would have been quite happy if another jolly (favourite) never passed the post. I spotted them standing at the bar as I entered, and Charlie, who was a natural at accents, was coming over with one of his endless supply of tales…

"So this Irishman said to the Father, it's a terrible thing Father, I can't stop gambling. Every day I must go into a betting shop and put money on a horse."

"That's bad my son, bad" said the Father. "You must light a candle and say three Hail Marys every day for the next six days."

Back went Paddy on the seventh day… "Father, every day I light a candle and say three Hail Marys but I am still backing losers."

"This is very strange my son, which candles did you light?"

"The small ones Father, I'm a bit short after losing all that money."

"Ah, that's the trouble then my son, the small ones are for the dogs."

# Chapter Eleven

The Goodwood of Greyhound Racing, is what Walthamstow Stadium has always been known as, and it is certainly very, very true.

Greyhound Racing at its most luxurious. Every possible comfort that could be provided for the racegoer is there.

The spacious lounges, with an all-round view of the track, are thickly carpeted, and the soft leather seats are a joy to sink into.

The track itself is very attractive, with illuminated coloured fountains and flower-beds, and an outsize clock saying how many minutes betting time is left before the start of the next race.

Flanking the northern end of the track is a register of all the bets being placed at the windows, inclusive of all the forecast combinations, which are such a popular bet at the 'Dogs' and at the southern end another illuminated board giving the up-to-the-minute odds.

The menu and service in the dining areas are comparable with tiptop hotels, and often better.

Unfortunately Johnny, George, Charlie and Bill were not to enjoy such exquisite surroundings, they were on the outside making a book, and working very hard at it too.

For over fifty years the Chandlers of 'The Stow' as it is popularly called, have been renowned for their Charity nights.

They have never been known to turn down a worthwhile cause, and regularly every November the Mayor, together with his Town Hall staff, could be seen going around with their buckets, making a collection of money for Christmas parcels for the Senior Citizens of the locality, and this night was no exception.

The Waltham Forest Band were there playing carols, not in their normal smart blue and black uniforms. but dressed in Edwardian style to fit the occasion, and the true Christmas spirit was rapidly descending upon the racegoers.

The lively rhythm of Sleighride coming over to George, made him wish he was on the other side of the fence with the Band, concentrating on getting his notes right on his trombone, rather than getting his figures right on the book. However, this was not to be, the money was coming for the first race, and they were beginning to get busy.

It was hectic, but after each race Johnny would relax a little, light a cigarette, and take things easy for a few moments, while George paid out the winners of the previous race.

## Chapter Eleven

It was during one of these relaxing moments that Johnny noticed a familiar pair of legs. One of the good things about Johnny was his memory. He could well remember a nice shapely pair of legs, as easily as he could remember the winner of the three o'clock at Newcastle three weeks previously, and the pair coming along now, to ask him for a contribution, definitely belonged to the little bird who came into his shop and put a hundred quid on Rose Petal The Third the day of the knockout.

Normally, he would mumble under his breath, 'on the bloody ear-ole again' and tell George to put a pound in the bucket, but tonight he wanted to make an impression... "George, a fiver for the lady – you are doing a very good job my dear". George, unable to believe that Johnny was throwing his money around like a man with no arms, was looking agape, but Johnny was far from being 'slow at the start' he was making a mental note that the Directors always entertained the Mayor and his staff to dinner in the Goodwood Lounge, and that he would have to make it up there a bit sharpish after the last race to chat her up, and get some information.

This was the first real lead that they had come across that remotely connected with the knockout, and not a bad lead at that.

It was going to he a pleasure chatting her up.

This, he thought, must be his lucky night, with most of the outsiders romping home as well.

He threw his cigarette away, and started to get into his stride for the rush of last minute bets on the next race.

As the minutes crept round on the big clock the betting really began to hot up, and for the two to three minutes before the 'off' Johnny was really going to town, grabbing the money from the punters, handing out the tickets, rubbing out and altering the prices on the hoard with Charlie tic-tacking over to the silver stand on the other side of the track He was going at such speed, and with such alacrity of movement, he would have put any self-respecting three-card trickster to shame.

After the last race Johnny rushed up the stairs to where the official party were preparing themselves to take their leave, so he lost no time in placing himself in the right position to make contact with the young lady concerned, and never at a loss for words started to lay on the charm. Much to his dismay, it wasn't working.

This was a situation he had never come across before, normally the birds chased him, and it was beginning to dent his ego a bit, but it didn't take him long to discover the reason for her coolness, it was in the shape of a young man, and from where Johnny was standing it looked like he knew how to handle himself.

"Oh well, you can't win 'em all."

At least he knew where she worked, at the Town Hall, and that she was local, they would all have to keep their eyes and ears open in future. Not that anything could be done officially, it was not a crime to beat the hook, only in the bookmakers language, and if they could get any information as to how it was set up, then they would have more chance of knowing how to hedge it in future.

The professional backers, the ones with the big money, were not normally involved in a knockout, they relied on their own skill and judgment, it was their living. The knockouts usually came about when somebody got hold of a good bit of inside information, usually when a horse is fit and in tip-top condition to race, and had the money to do something about it.

The bookies had yet to tumble how the knockout was done, there are various ways, but in this particular instance the market racket had been worked.

As everyone knows there are wholesale fish and vegetable markets in every principal town in the country, and they all make an early start to the day and finish work around noon.

About fifteen to twenty leaders are selected, who work in the markets, and who can, to a certain extent, be trusted, in towns all over the country, Manchester, Liverpool, London, Newcastle, Birmingham etc. etc., and either £1,000, £2,000 or £5,000 is made available to each Leader at very short notice before the race. They in turn each select ten men who are working in the market and who live in highly populated areas, where there are at least ten betting shops within easy reach of each other, and the men must know how to get from one shop to the next pronto, and even more important, how to keep their mouths shut.

The main men at the markets are contacted approximately one hour before the race, and given the name of the horse to back. They divide the money they have been given between the ten men they have selected to back the horse in ten individual bets at their ten selected shops at the latest possible moment before the "off".

Now with twenty markets placing a possible £1,000, £2,000, £5,000 or even £10,000 each around the country, that is a possible £200,000 going on in £100 bets without giving the bookmakers time to realise what is happening and to hedge the bets and therefore, bring down the price.

The boys placing the bets are usually given £10 each to put on for themselves, to keep them sweet for the next fix, so that if the horse comes up with an S.P. of 4/1 they come away with £50 for about ten minutes work, and they normally put another fiver on for the missus. The technology used today is so sophisticated that other methods have to be worked out.

# Chapter Eleven

However, all was not lost. On the Thursday it was Charity Night at the local Room at the Top. Many local businesses were giving their support, and some well-known artists were giving their services free. It was in aid of the childrens' fund, and a few members of the band who were there to play the introductions had persuaded me to go along. I had bought a £15 ticket anyway, but thought I would feel a bit out of place on my own, hut I could hardly refuse when they said they were bringing it up at the next Committee Meeting of the Band that they should ask me to accept the Vice Presidency, which had been unfilled since the death of Ken. They appreciated that I always turned up at the functions at which they played, and cheered them on at the various contests they entered.

Johnny was looking more handsome than ever in his evening attire. The local betting shop making a generous contribution to the evening.

Pity nobody thought to ask Charlie to partake in the entertainment, he could have kept them amused for hours with his string of tales. He was where he could usually be found at the bar, with a crowd round him, as usual, all waiting for the latest...

"...and so this Rabbi said to his assistant, 'now go down to the Catholic Church and see why they should do more business than we do'. Off went Ivor, and was back within half an hour... 'vor vhy are you back so soon'... because I know vhy they are so busy, no sooner than they are in there when a geezer comes round with a plate to collect the bets... sure that's why they're so busy..."

He called the barman over for a refill and went on...

"and this same Rabbi was even more annoyed when the Priest, who lived opposite him went out and bought a new Mini, brought it home and baptised it by sprinkling some Holy Water on the bonnet. Off rushed the Rabbi purchased a Jag and came back and chopped a couple of inches off the exhaust."

l am sure Charlie could have kept it up all night, it was the way he told 'em, but it was time for us to take our seats for the dinner, and Johnny could not believe his eyes when he saw that to be seated right there opposite him was the girl with the smashing legs, and she was minus the boyfriend, and was being escorted by her old man who owned a shop in the High Street.

It didn't take Johnny long, when the dinner and speeches were over, to become more closely acquainted with her. He knew how to put on the charm, and at the same time completely confusing her with talk about winners, prices, and asking her opinion about various performances.

She was completely flummoxed "I'm sorry… I don't know anything at all about horses and racing. I have never been to a racecourse in my life."

Johnny was really laying it on thick… "Oh come on, you don't expect me to believe that do you. I remember very well one day, when you came into my shop and put a ton on a horse that won easily at 4/1. You have got to know about racing to come up with something like that."

She was most indignant "…that wasn't for me, that was for my boyfriend. I only ever have 50p on the Derby and Grand National."

Johnny went on "…if you'll pardon me saying so, your boyfriend must have a very good job to be able to put that sort of money on a horse." She was really pouting. "My Brian wouldn't put a pound on a horse, let alone £100."

This, Johnny could well believe, from what he had seen of Brian he looked a right skinflint, not at all the type to contribute to the expensive upkeep of a bookmakers life-style.

"We are saving up to be married. My Brian wouldn't throw his money away like that. He put it on for a friend where he works at the fish market. He paid him, and as it was legal and we need all we can get towards our mortgage, he agreed to put ten bets on for him, but Brian's car was playing up that day, and he was afraid he wouldn't be able to get to all the shops in time, and as I had a day's leave, we were going to look at some furniture that afternoon, I said I would help him… did it win?"

I glanced over at Johnny, I could imagine what was going through his mind – oh cor blimey, did it win. It makes your heart bleed, but at least he now knew how it was being set up. The next thing would be to find out who was setting it up. It had to be someone who could afford it, and was copping hold of some good information, and also had connections in the markets.

Not much to go on, but should narrow the field down a bit. He made a mental note to send a memo to his other shops to keep an eye open for last minute betting in tons, and to hedge straight away, whilst there was still time.

Once this sort of thing started it was inclined to roll.

Some bookmakers decided to get their own back by keeping the prices down at the afternoon dogs to about 2/1 at the track hedging a load of bets, and then putting up the prices to about 7/1 just before the off to get a good return price on their hedge bets.

## Chapter Eleven

As I opened the front door, I could see the light flashing on the telephone in the hall. I just managed to get to it as Steve was about to hang up. "Hello Val, had a good evening?" "Super, and they raised about £12,000 on the night."

"Good show"

He went on "…I wondered if you have seen tonight's paper. Another horse has died, in rather similar circumstances. Not one of Amanda's this time."

"There seem to have been a few horses dying in the same way over the past year, I thought you might like to check, see if they happened to be insured by the same person. At the same time you might see if anything else connects. I noticed myself, that the last three horses to die have all been past their prime, and towards the end of their racing career. May just be a coincidence of course. We will have to get together on this Val, two heads have got to be better than one." "Give me a ring tomorrow and let me know if you come up with anything. Goodnight my dear, sleep well."

I knew he meant well, but was at a loss to understand how he could say 'sleep well' after filling my head with a load of questions. Surely I would be all night trying to figure out the answers, but as it happened, after such a tiring day, as soon as my head hit the pillow I was asleep.

# Chapter Twelve

On my way to Church the following Sunday, I made up my mind to have another word with "Our Lady" and to ask her to pray for me for help to untangle the mystery of Ken's accident (or as I had come to believe) his murder.

I was determined to get to the bottom of it – I would never ever accept the Coroner's verdict.

The Church was as beautiful as ever and the Priest looked superb in the new vestments someone had forked out for, but there was a problem, everyone looking very glum, there had been a very high wind during the week and the roof needed some immediate repair if further damage to the Church was to be avoided.

Everyone was thinking up ways to raise some money, but it needed quite a lot and quickly.

So back to the dog track.

I had got to know Percy – the boss – through the

various charities I had supported there, and so I went to see him and begged him to let the choir do some Christmas Carols and songs to raise some money for the roof.

I could see what he was thinking 'dog racing and church choirs were not a good mix,' but I explained that we would dress up in Christmassy gear and bring our gold outside candles with us which are very decorative, and we would keep to well-known Christmas songs and Carols, such as White Christmas and The Red Nosed Reindeer.

He was not really convinced, but after thinking it over he said they always have a free night just before Christmas – no entrance fee – and he thought that some people may decide to give, as they did not have to pay to go in.

He suggested we stood behind the turnstiles at the main entrance and hope for the best.

The lads and lasses of the choir did not have time to rig up Christmassy gear, so they took along with them cassocks, mitres, choir robes and anything else they could think of to make a bit of a show.

Percy came down before racing began to see how we were getting on. He was absolutely gobsmacked to see the money pouring into the buckets, and not only that but many people were standing actually listening, and not rushing through to get their money on the first race.

He seemed genuinely surprised that the choir were so good, and there and then asked me to ask them if they would stay after racing and sing some popular songs in the Goodwood lounge.

## Chapter Twelve

"I am sure they would love to Percy, but they have all come here straight from work and have not eaten."

"No problem, I will arrange for food and drinks for them on the house."

"What all twenty-four of them?"

"Of course, and we will also make a generous donation."

He arranged for the money we had collected and our gold candles to be taken to the strong room, and I was amazed to see a fantastic machine that you just pour the money into and it sorts itself out. The staff in there did this for us and it was all sorted within a matter of minutes, and they then exchanged it for notes, which made the going home so much easier, rather than carrying the heavy buckets, and also safer.

I had only taken a fiver with me, as I did not want to be lumbered with a handbag – and as the boys and girls said they had no money, I put the five pound note in the centre of the table and suggested we took it in turns to select a dog and put fifty pence on so that we had an interest for the eight races.

Just before the seventh race Percy whispered in my ear...

"If your friends fancied a bet I think the number four dog should do quite well."

Miraculously they all managed to find five and ten pound notes, which they didn't have before, and of course the number four dog walked home.

The following week I was reading the local paper, as I had taken an interest in it since the phony reporter had turned up at Simon's yard, and what a giggle I had when I saw the headline on the Sports Page –

*WHAT A JOY TO SEE ALL THE CLERGY LINING UP FOR THEIR WINNINGS AFTER THE 7th RACE AT WALTHAMSTOW LAST THURSDAY EVENING*

# Chapter Thirteen

More problems, because I didn't have enough already, my neighbour was utterly and totally the biggest snob I have ever known – friend of Lady Tuckington you know – Lady This, Lady That and Lady Muck. She thought she was slumming having to live next door to me, poor soul, and a polite good morning or good evening was all that was ever exchanged between us.

One day her husband, a very smart man who regularly had his hair permed, was having morning coffee and a pastry, when he keeled over with a heart attack and was dead before he hit the floor. What I did not know at the time was that Her Ladyship was alcoholic. She found the key to the liqueur cupboard before he drew his last breath. It was a steep downhill road from then on. As soon as the Undertakers took him away, she locked the front door and made inroads to whatever was in the cupboard.

When she finally ran out of booze she opened the door to a gentleman who had been trying for a week to get an answer. The lady who always looked so ravishing when she went to Ascot or Henley with her husband, looked like something out of a Frankenstein film. Her beautiful negligee had not been taken off for over a week her very fine hair which was normally in a bun had come adrift and was hanging loose over her face which was a good thing in some ways because it hid what was underneath.

The gentleman, I found out afterwards, was a friend of the husband who had promised to look after her should anything happen to him. The funeral was obviously arranged by him, and he and another two gentlemen spent some time banging on the door trying to arouse her – they finally got her out of the house by a strong arm either side and dumped her in the car, following the hearse.

A few days after the funeral this very imposing-looking gentleman called on me and asked me if I would keep an eye on her. I told him that I work full-time and did not want to be responsible. He said if he left me fifty pounds would I get her some food, and he was also leaving a cheque book for her to pay the electric, gas, rates etc.

She had never had a cheque book before because although her husband opened accounts for her to buy clothes etc., and she did have beautiful things, he would never let her have money, which was understandable, but she managed to find a little corner shop run by an Asian family, and persuaded the man to let her have

## Chapter Thirteen

whatever drink she wanted and he would make out the cheques.

I had been told that there was no shortage of money and more than enough for her to live very comfortably for the rest of her life, so I was very surprised one day when she asked me to lend her a teabag. She did this on a regular basis knocking on all the doors asking for a teabag and a slice of bread.

Why don't you buy teabags and bread I asked, obviously the fifty pounds worth of groceries I had got for her had run out, "I haven't got any money".

"Of course you have, you have a lot of money"

"Oh no" she said "I have come to the end of my cheque book", and she thought that it meant that she had come to the end of her money.

Two of the ladies who she had previously played bridge with thought that they would be kindhearted and invited her to join the Guild of Friendship of which I was a member, the only thing was I got lumbered with her.

Every year we used to visit a different Cathedral, and the one we had chosen for the current year was having a very special celebration and the Cathedral had the most beautiful flower display and there was to be a musical celebration all the choirs from the surrounding areas were to take part and a platform had been built together with stands to accommodate the visiting choirs. There was a rostrum for the Conductor in the central part of the Cathedral.

Camilla, as she had started to call herself, decided she would join us on the trip. She was well away before we started. My friends were all saying "don't let it spoil your day" but as they made sure that the only seat available to her on the coach was the one next to me, I had to put up with her, and I couldn't see how I could not let it spoil my day.

Whenever we stopped at a Service station I was buying her pots of tea, she would not drink coffee, to try and sober her up. It was not working. When we arrived we were having lunch that had been arranged prior to our visit, and I saw the opportunity to dump her on her two bridge friends when I spotted a priest I knew who was a friend of a well-known trainer – and I wanted to make contact and have a chat with him.

I got away sharpish when the meal was over, and thought that I would be free for the rest of the afternoon. How wrong can you get. I was in the Ladies Cloakroom when I bumped into the two friends.

"Where's Camilla?"

"We don't have anything to do with her any more."

I might have guessed. "What happened?"

These two were almost as snobbish as Camilla used to be, but could not resist telling me the dreadful news.

"Fiona was having a sherry morning for the Conservatives, and Camilla was the first to arrive. Fiona had poured eighteen sherries, six sweet, six medium and six dry." She looked the tight-fisted type who would only splash out on three bottles. Emily with a terrible frown went on "and when Fiona popped into the kitchen to get the nibbles, she came back to find the eighteen glasses empty, and Camilla flat out on the settee."

## Chapter Thirteen

That was the best part of my day, I hooted with laughter which didn't last long because when I came out of the cloakroom and into the main part of the Cathedral, there was a bit of a rumpus going on on the rostrum – Camilla, still not sobered up, was up there trying to conduct the choir.

The Stewards got her down, and I thought it best to get her out of there. We went to the front entrance, she fell down the stone steps and pulled me with her. Fortunately I fell on top of her, but she had badly cut her legs and hands on the gravel. The first-aid men arrived with a wheelchair and we went off to the first-aid room. I hadn't realised until then just how much I was shaking, but it was so nice and warm in there and they put a blanket around my shoulders and made a nice cup of tea. They patched Camilla up with bandages and plasters but very strongly advised her to go to hospital, but she would not have it, however badly she hurt herself she would never see a doctor or go to Casualty, she knew they would try to advise about her drinking and she didn't want any advice, the only advise she wanted was to be told how to get hold of it. It was most disturbing for me, particularly in the winter when it was dark about four o'clock, she didn't know if it was afternoon or night-time and very often was ringing my front door bell which starts the flashing lights going at three to four in the morning. This was making me extremely nervous and with all the other strange things that were happening I was losing confidence. My hearing deteriorated rapidly and I became profoundly deaf. Things were not getting better, in fact they got worse. I was jumping at every movement, and having to adjust to a much more silent world.

It was then suggested that I should apply for a Hearing Dog For Deaf People. I was a little unsure, I did not want to be too restricted, but I was assured that Hearing Dogs were welcome in most places, so the very first thing I did before I decided was to write to all my favourite racecourses to ask if I would be able to take my dog on race days.

The replies gave me a terrific lift. They all said that a Hearing Dog would be very welcome. I made up my mind immediately, and sent off for an application form.

The first thing is an audiological test, you have to be severely, profoundly or totally deaf before you are considered, then Gill a Placement Officer came home for an assessment, see what my dog needed to be trained for.

Once the paperwork is in place, a suitable dog is put with a Socialiser, same as a Puppy Walker, just a different name, someone who gets the dog going on general obedience, takes it to the park everyday, teaches it to heel at kerbs etc. When the dog is a suitable age, normally about ten months old, it is taken to the Training Centre to train for the work it is going to do with the recipient. This normally takes about four to five months. The trainers are firm but kind, they are so patient and really love all the dogs, and the dogs are rewarded with a small treat when they get it right.

The first thing they learn is to alert, by tapping the knee or jumping up to the hips, no good barking to get attention, the recipient wouldn't hear. Once they have got the recipient's attention they lead him or her to the source of the sound. Morning alarm, door bell, cooker timer, telephone for when a fax is coming through, smoke alarm, fetch, and if necessary a baby crying.

A little later on I got a letter with a photograph to say that a dog had been chosen for me and would I go to the Training Centre to meet the dog – see if we were compatible, because if we weren't it would not work.

I was very excited, couldn't wait to get there, and luckily we hit it off straight away. So now the training could begin in earnest.

# Chapter Fourteen

In the meantime I had made up my mind that I really must go through the files again of money paid out for horses that had died rather suddenly, but I had to get myself really fit to be able to do this satisfactorily.

Christmas was a-coming and I thought it best to start my investigations in the New Year – start afresh. I was looking forward to Kempton on Boxing Day for the King George. For as along as I can remember, I had gone to Kempton on Boxing Day. Many years ago we used to take a hamper of turkey sandwiches, a bottle of wine and pay 3/6d. seventeen and half pence in today's money to go up on the Terrace, and freeze for the rest of the afternoon.

It is a very different story today. £185.00 for the day at the Jubilee Club, but well worth every penny. I always book from one year to the next as it is so popular, and have the same table near the tote and near the bar so that everything is to hand without too much disturbance, coffee and pastries on arrival, a fabulous lunch and tea in the afternoon, and hopefully a few winners. It has a bit of a carnival or Christmassy atmosphere about it and everyone seems to be in a good mood punters and staff alike.

## Chapter Fourteen

New Year, and Johnny, Charlie, George and Jimmy were preparing themselves for Cheltenham.

Charlie had taken his kids to Devon during the previous autumn and on the way home they had passed a farm with a notice outside saying 'TEAS'. The children were getting restless with the long drive and begged Charlie to stop so that they could have tea and see the animals.

The dear old lady who owned the farm, a motherly type, with snow white hair a lovely complexion, obviously she had never used make-up to spoil it, and hands that looked as though they had worked hard all her life, made them very welcome, saying the kids could go and see the animals while she prepared the tea. She told Charlie and his wife that her husband had died and that she could only afford to keep one man on to help her run the farm, and that it had always been her home, previously belonged to her father, and she didn't want to leave, but was at her wits end to know how to keep it going.

It suddenly occurred to Charlie that it was not too far away from Cheltenham, and that it was very difficult to find suitable accommodation when the racing was on. This would suit us very nicely he thought, so he gave the old lady a very good tip, he was a generous man, and suggested to her that if she would allow him and his three buddies to stay during the racing at Cheltenham, they would look after her very well. All they would need would be a fry-up for breakfast and a meal when they returned in the evening. "We won't be any trouble at all" he said we will probably play cards for most of the night and finish off sleeping in the chairs, which were very old fashioned but quite comfortable, "in all probability we won't use the beds at all much."

She was very reluctant, "I've never done anything like this in my life before, I don't know if I could do it."

"Of course, you can," and the figure he suggested they paid her convinced her it would be worth her while and certainly help towards paying off some of the outstanding bills, which ultimately would save her a lot of worry.

He left her a deposit, and couldn't get home quick enough to tell Johnny George and Jimmy how clever he had been.

When the time arrived they packed all the booze they needed into the boot of Johnny's Merc.

Charlie as thoughtful as ever "Better put a bottle of sherry in for the old lady".

Mrs. Hill had a lovely meal ready for them when they arrived, and a lovely log fire going in the drawing room, as she called it. There was a big oval table in the centre of the room, and a side-board (or what used to be called a chiffonier in the old days) against the wall. Very accommodating for all the drink. They asked her if it would be o.k. to put the drink out and she found a lovely old cloth to stand it on.

"We've put a bottle of sherry in for you dear" said Charlie.

"Oh no, I never touch it – Chapel you know."

"Well help yourself to anything you want, there's plenty there".

They played cards all night, and after the most

## Chapter Fourteen

wonderful fry-up ever they left for Cheltenham.

When they returned from the first day's racing, the smell of the dinner to come and the warmth of the log fire, and of course most of the favourites being beaten – things couldn't get better.

Jimmy, the youngest of the four and was the runner for the bookie, noticed that the sherry bottle had been opened and had gone down. "Crafty old cow, her and her Chapel – she likes a drop".

Charlie was furious, he had really taken to the old lady.

"Leave it Jimmy, I told her to help herself, we brought it for her we won't drink sherry."

On their return after the second day's racing, Jimmy couldn't wait to get in to see if the sherry had gone down again, and it had, there was less than half a bottle left. Jimmy Hall was dancing up and down "I told you, I told you, she's at it, good job it isn't the scotch she's having a go at."

Charlie was getting really mad with him.

"Will you leave it, it's not doing you any harm, forget it" but would he no.

When he wanted a pee he decided to top the bottle up. Serves her right it's hypocrisy going on about her Chapel.

The time came for them to leave, they had had a good Cheltenham, and all (but Jimmy) put in a very good tip for the old lady. They were packing up what was left of the booze and noticed that the sherry bottle was empty.

"Come on darling, come and have a little drink with us before we go."

"No sir, I told you I never touch it, Chapel you know."

"Not that we mind at all, but we couldn't help noticing that the sherry went."

"Well Sir, the first night you came Mr. Hall said how much he enjoyed my trifles, so I have been putting the sherry in his special trifle every day – you did say I could help myself."

## Chapter Fourteen

Johnny and his staff had had a very good Cheltenham. Most of the favourites had been beaten and they really enjoyed staying at the farm, but there was calamity at the betting shop when they got back.

Old Albert who made it his second home, by bringing his little pack of sandwiches each day and paying the girls 35p to make him and cup of tea or coffee from time to time, sat in his little corner every day from opening time until they closed. He was quite harmless and everyone was used to seeing him around, but he would give tips. They never came up of course, and nobody took any notice, but on this particular day he was giving anybody that would listen to him a tip for a horse that they were laying at 7/1. The race began and everyone had their eyes glued to the television screens, and quite unbelievably the horse got a flyer at the start and was about five to six lengths in front it looked as though nothing was going to catch it, when it broke a blood vessel (it happens from time to time) and died, and so did old Albert, with shock.

An ambulance was called and he was dead on arrival at the Hospital. The next day, the Police came into the betting shop to say that they had been to Albert's flat and asked his neighbours, but everyone was of the opinion that he had no living relatives, and he had to be identified. What they had come to ask was that as he spent best part of his life there, would a couple of the men identify him. They put their heads together.

"He wasn't a bad old boy, although a terrible tipster, and the least they could do was this last service for him."

So off they went to the mortuary, and it was most unfortunate that there was a new mortician's assistant that day who had yet to learn the ropes.

He took them along a corridor and into a room and pulled out the first drawer, gently lowering the sheet.

"That's not Albert" they said in harmony.

"Sorry sir."

He pulled out the next drawer and repeated the action of gently lowering the sheet.

"That's not old Albert."

"I must apologise it's my first day here, I will try again."

The third draw was the same.

Well it's not old Albert, but it's absolutely typical, he never was in the first three.

# Chapter Fifteen

It was a miserable, cold, depressing day, very overcast and drizzly when I bumped into Philip at Lingfield. I hardly recognised him, he looked ten years older, his hair was turning grey, and he had lost a lot of weight.

"Hi Philip, how are you?"

"Not bad Val, not bad."

But I could see that he was, he was just a shadow of his former self.

"Look Val, we can't talk here, but I would like to have a word with you, a lot has been happening recently, and, with hindsight, I believe that Ken was on to it, or at least some of it. He knew a lot of people in the racing world, both through his work, and the love of his hobby, and it wouldn't surprise me if somebody had spoken to him about the strange events of the last year or two."

"Phil, I have been certain from day one that Ken's car crash was no accident, but what could I do, no one would believe me, there was no other vehicle involved, and once the Coroner said Accidental Death, due to loss of concentration, I was on my own, apart from Steve who has been very good in trying to help, but we have not got very far. The police deal with many accidents every day and made it quite clear to me that they were not prepared to investigate further."

"How about coming and having a bit of dinner with me one evening when you are passing my way?"

I was about to say, "not Tuesday or Thursday because of my Judo class" but thought better of it, it was more important to hear what Phil had to say than to miss an evening at the class.

"Would tonight be too inconvenient, I will be passing on my way home, and I really want to get this off my chest?"

"Not if you don't mind a take-away, I won't have time to cook."

"No, that's fine, and I will bring the wine."

I don't know if it was the miserable day, or learning what Phil had to say, but I couldn't seem to concentrate on finding winners. It was a bad losing day. It reminded me of years ago when I had a bad day and my family were with me, they used to sing to me "There's a Pawnshop round the corner in Pittsburgh, Pennsylvania".

No pawnshop but credit cards today and I would have to resort to mine for the take-away.

Kayla made herself comfortable in the car on the way home, her first day's racing, and she really enjoyed it. So many people made such a big fuss of her.

## Chapter Fifteen

I made sure it was really warm and bright for Philip when he arrived, he certainly needed cheering up.

He hit it off with Kayla straight away, and I think he found some comfort in cuddling up to her and stroking her, and of course she was loving it.

The Indian meal and the wine made us both more relaxed, I was a bit bothered about what I was going to hear, and how I could deal with it, and we left it until we got to the coffee, very difficult to lip-read when someone is eating. Go into a restaurant or cafe where a deaf group or club is having an outing, and there will be complete silence during the meal.

"The problem is Val, that that silly old fool Saint Paul has got himself into severe trouble through his do-good visits to prisons, trying to save the hard-done-by inmates. It was a time-bomb waiting to go off. It was a racing-cert that one day an old lag or a group of old lags would get to him, and now it has happened. He is being blackmailed, digging himself in deeper all the time, and dragging me with him."

"As you know, I have worked for him for many years, and on the whole I have been happy in my job, but these last months have been a nightmare."

"One of the prisoners in a jail that he visited is a member of an east end gang who have for a long time run a bit of a protection racket at the east-end markets. Nothing like the Mafia, of course. very small scale really, just throwing ink over the clothing stalls or tipping over the fruit on the fruit stalls, things like that if they didn't get their weekly dosh. Well it's coming to an end, the stall-holders are getting together and trying to protect each other by forming groups and standing-up for themselves on payment day, so the Chopper gang had to think up a new get-rich-quick scheme, and Saint Paul provided them with the opportunity.

"It was well known that he knew many people in racing circles, and he was seen at the races most days, chatting to head lads, trainers, jockeys, in fact anyone who would listen to him, and not many would if they could help it.

"Chopper and his mates had a meet, and decided that Saint Paul would do what he was told or the horses would suffer.

"And this of course is what has been happening, particularly to Amanda's horses.

"Ragwort has been found in the hay, and acorns have been spread about where the horses graze, both of which can be quite poisonous. They are now getting at him to try and bribe certain stable lads and lasses, but of course this is not possible. The lads and lasses love their jobs and the horses they look after, they wouldn't do it if they didn't love the life because it is such hard work. As you know it's early morning start and they really don't have much chance of a social life, the bonus being when they take the horse to the course to race for the day."

Philip went on… "That he had really worked hard for Saint Paul over the years, protected him from many misfortunes and kept all the facts and figures relating to the many companies he had an interest in, in apple-pie order".

"In fact, Val, I've spoilt him, just like spoiling a child, and he now thinks that I can work miracles for him, get

## Chapter Fifteen

this gang off his back, but it's not that easy, Chopper and his lot think they have taken over from the Krays, but it's a very different world today, it's the druggies and Yardies who have the monopoly and Chopper's gang are very small fry in comparison, but they obviously have something on St. Paul to be able to blackmail him the way they are.

"You will remember when I told you about Amanda finding the drugs in the pad of the shoe of the horse sent to France, well that was only a very small part of a much larger operation that Saint Paul has got himself involved in, I cannot think why, it is not as though there is any shortage of money and that he desperately needs some, it probably came about in the first place because he is so weak he would agree to anything if he thought it would make him popular, and I have always suspected that his friend and co-owner is a bit dodgy. I have saved St. Paul in the past from some unscrupulous deals with him, and I have been sure for some time that he is only using Paul for his own ends, but this time I didn't know anything about it until it was too late, he was already heavily involved.

"Of course there has to be an outlet on the streets for drugs and this is where Chopper and his mates came in with their market connections. They must have learned about Paul's indiscretions and are now playing on it.

"I agree with you Val, I think Ken knew something about all this, and that probably they did not intend to kill him, I think they were giving him a warning to keep quiet, which of course was ridiculous because a man of his integrity would not be put off."

"The question is where do we go from here?"

"I have done my best but now I feel I would just like to resign and escape somewhere to get away from it all, but of course I have to live and it would be difficult for me to find another job like the one I have got at my age."

We mulled it over together...

"What about if I had a word with Steve, and put him in the picture about what you have told me tonight. He uses an agent from time to time, and even suggested some time ago, that we appoint him, but at the time I felt we had so little to go on, but now we have really got something to give him to work on and make a start, I think we could possibly get somewhere with it."

"This is a fantastic idea Val, but you know agents can be very expensive and can run their expenses up to enormous amounts, I really cannot afford that."

"Don't worry about that, Steve and I will attend to that side, you just keep telling us all you know as it comes along."

I knew Steve would be willing and I felt sure that the firm would agree to cover the cost if it meant that Mike could unravel this mystery.

I faxed Steve to arrange to meet him the next day.

He was both surprised and interested in what I had to tell him, and agreed straight away that he should contact Mike and give him all the facts we knew to get him going on this assignment. Philip was so relieved that at last he was getting some help, and felt that a very heavy burden had been lifted from his shoulders.

## Chapter Fifteen

A grin slowly spread over Mike's face.

"Oh yes, Mr. Stevenson, I know the gentleman you are referring to, and his mates. As a matter of fact, without breaking any clients' confidentiality, I am working on something about him at the moment."

Steve was all ears.

"You probably know he has been running a protection racket, in a small way, at the east end markets for years, every Friday he and his heavies, who are not at the time having a holiday at Her Majesty's Pleasure, go round to collect the donations or contributions, as they like to call them, and the stallholders have in the past paid up to keep the peace, they always thought "I shall just have to sell one more jacket or dress, or a few more bits of jewellery, or another box of apples and pears to pay for it," but today the life of the stallholder is not as easy as it was years ago, then they could put their stall out once a week, mostly on market day, or if they were very ambitious, Thursday, Friday and Saturday, and make a very good living, but with most of the High Streets now having superstores and major stores with competitive prices, and of course the big boot sales, the stallholders are having to work doubly hard, going to more markets during the week and it's not really an easy life having to get up early, set up the stall and possibly stand there all day, very often when it's pouring with rain, or freezing cold in the winter.

"They have to pay the Councils for the pitches, and are now reluctant to pay Chopper and his mates for what he calls insurance.

"There have been one or two skirmishes recently with those who have refused, but they really wanted to put an end to it for ever, so they approached me to advise them how to go about it, and I have been working with them recently, but I know nothing about the problem with the horses."

"I will get on to it straight away, Mr. Stevenson. It will be the usual fee, and expenses, and I will send you reports just as soon as I have made inroads and there is something worthwhile reporting."

"Anything else you can give me while I am here?"

"Afraid not, it seems people are being very legitimate at the moment."

# Chapter Sixteen

The Central Line as usual was playing up, and I was getting uptight as to whether I would get to my meeting in time at the City Lit.

It was being held for people who had become deaf quite suddenly and were finding it difficult to cope. I had been warned sometime ago that my hearing would not get better, only worse, so I joined a signing class to learn British Sign Language, or I preferred Sign Supported English. I was not particularly good at it but it helped me to get by from time to time.

At the meeting they had Palanatype, which was on a large screen, a lip-speaker and an interpreter of sign language, but even with all this I was finding it difficult to follow.

When we broke up for coffee, I got into conversation with a gentleman who was a Doctor – Doctor Green. He was not much to look at, short, skinny and a wrinkly face for a man of forty-two, but he had a wonderful sense of humour. He told me that he had tripped up a kerb one day when getting out of his car, knocked his head against a concrete lamp-post, and hadn't heard a thing since.

Very sad for professional people who lose their hearing suddenly in middle age, they have usually established themselves in a good job, with a family and a mortgage and they are unable to carry on with their profession. I have seen it happen to Teachers, who have had to give up their jobs.

Dr. Green told me that he manages quite well with a very efficient secretary who had been with him for years, and knew the routine and general running of the surgery inside out. He went on… "But of course she has to have holidays from time to time, and that's when things started to go wrong".

Only the previous week when he had a temp in to do her job while she was away, an embarrassing incident had occurred. He got to the Surgery after making his afternoon home visits, where the temp was in a bit of a panic. She had taken a call from Mrs. Hill in Carisbrooke Road who was haemorrhaging very badly and would he go to see her immediately. Now his own Secretary would have known that there are two Mrs. Hill's in Carisbrooke Road, one the dear old lady of ninety-one who was poorly, and her Grand-daughter-in-law who lived a few doors down who was twenty-three years old, blonde, and very curvaceous. His own secretary would have known immediately what records to call up for the telephone number, but of course the temp got the wrong one, the phone number of the young one, and as he was rushing out of the door, he called out "Tell her to leave the front door open, go upstairs, undress and lie on the bed, I will be round in a few minutes."

The meeting was helpful in many ways, but I wondered if I was ever going to adapt to this silent world.

I was advised to give up judo, it was getting a bit too much for me anyway, and instead I decided to join a signing choir. We sign hymns at church services, popular songs at groups and clubs, such as Womens' Institutes, Townswomens' Guilds, and Senior Citizen Clubs,where they liked all the old tunes, such as Daisy Daisy, Sally, (our Gracie's song) When You're Smiling and songs from the shows, so that they could sing with the musical background. We went to Jersey and signed Hymns in the Church there, and we also signed Carols at Shopping Centres, Lakeside, Liberty 2 and the Exchange to raise money for helping the deaf and hearing impaired. We really enjoyed it and it helped us communicate with hearing people, particularly when we took our Hearing Dogs with us, they certainly attracted a lot of attention.

The following Sunday, the Band had entered for the South of England competition and invited me to go along with them to Brighton. We had to make an early start because the competition began at ten o'clock, and they liked to get there early to change into their uniforms and get generally settled down.

I popped a few bottles of Bucks Fizz and plastic wine glasses into a cool bag in case there would be a need for celebration on the way home. They usually played well, and managed to capture some prize.

The competitions were held in a very large hall where a sort of open-topped tent had been erected in the centre. The band had to play inside the tent so that the Judges could not see who was playing and relied on their marking solely by the sound.

I went off to find a nice little church, as there was little point in my staying. The familiar smell of incense as I entered made me feel quite at home. It was a lovely Service, all smells and bells as they like to say in the High Church, and they made me very welcome. After the Hail Mary, Full of Grace, we went to the back of the church for coffee and biscuits and they were so friendly and nice, I thought wouldn't it be truly heaven if life was always like this, no selfish and horrible men beating up and making other peoples lives unbearable through jealousy and greed. It was lovely to get away for the day, away from the worries of the past few months, and there was even better to come.

I strolled back to the Concert Hall to see how the band was getting on. They had played and were free to go to lunch. Bill and five other members had booked a

## Chapter Sixteen

lunch at a small private hotel, and they asked me to go with them.

Brighton was at its best, and the weather was being kind, quite warm and bracing. I was beginning to feel like I was on my holidays, as we walked along the prom taking in the good sea air. The men were in high spirits as they felt they had played really well that morning, so it was turning out to be a very good day.

It was an old-fashioned family run hotel, which was spotlessly clean and good home cooking which smelt delicious. They served the vegetables in small stainless steel dishes. Bill was sitting next to me and urging me to have some courgettes, "You must have some Val they're absolutely lovely". They looked liked carrots to me, a veg that I disliked, but not to make a fuss I took some. They tasted like carrots. What a mystery. However, this was solved the following Tuesday when I met Bill's wife. She told me that now the competition was over, they were going on holiday. Bill insisted on driving but she was unhappy about this because he was colour blind and found it difficult to follow the traffic lights.

Now I know why I was eating carrots.

The Band came second, which was very good, and the atmosphere in the coach going home was terrific.

# Chapter Seventeen

Mick knew the East End like the back of his hand. He had many connections he could call on and also his own snouts, who he paid money to for information.

He was sitting in a café in Hackney, consulting his little black book trying to work out who would be the best to approach about the new Chopper interests. Who better than Charlie at the bookies, he knew about everybody and his information went back a long way. His father was born in Hoxton and he was in his prime at the height of the Krays and worked for them from time to time as a sort of odd job man, running their errands and run-of-the-mill jobs, he was not into the violent side, but saw plenty of it.

Charlie was always at the races with Johnny making a book, he was his chief clerk, and was an expert at figures, and could have made a very good accountant if he had had the opportunity to go to college when he was younger.

He had seen too much heartache when he was young as a result of the war of the gangs his father worked for, and although he liked mixing with those on the other side of the fence, there was always something interesting going on, he was determined to go straight himself. He loved his family and would guard them very carefully to make sure no harm came to them, but on the other hand because of his unique position he was able to pick up a lot of good tips which he passed on to Mick from time to time.

Charlie was propping up the bar at the Duke's Head as usual when Mike found him.

"Wotch-yer mate how's it going."

"O.k. and as I get the feeling that you are going to bend my ear, make it a scotch, a double".

"Let's go and sit over by the window so we can dodge Patsy joining in every two minutes."

Patsy is a good barmaid, but didn't like to miss anything, was always aware of what was going on around her. They picked up their drinks and made their way towards a table in the corner, where it would be difficult for anyone to overhear.

Mike dived straight in…

"Chopper, a couple of his oppos seem to be missing."

"Yeah," Boots who got his name because he has been wearing the same winkle-picker boots since he was a Teddy Boy in the sixties, "He's in the Scrubs, went down for G.B.H. when one of their 'clients' refused to cough up."

"Don't know much about the other one, they sent him to a jail up North."

## Chapter Seventeen

"Sticks, who has just got out, has been laughing his head off since he got home, telling everyone that that barmy old fool St. Paul, is trying to turn him into a born-again Christian, and he's playing up to it, saying he's going to turn over a new leaf. Some hopes."

What a break, Mick could see why he wanted to play up to him, so that he would keep visiting and Boots could put more and more pressure on him about the horses. It was such a waste of time, it was never going to go anywhere beyond the initial stages, but they really believed that he could one way or another fix races.

They couldn't see that St. Paul wasn't clever enough and that he was just fumbling about, getting himself, Philip and Amanda in a right old state, he really thought that if he supplied Philip with the necessary, Philip would actually bribe the stable-lads and lasses, which of course was out of the question. Philip knew that the technology of today was so far advanced that there wasn't a hope in hell of putting it into practice.

However, in the meantime they carried out their threat of more horses dying, and it was becoming increasingly obvious that this wasn't just bad luck, and that the feed was being tampered with.

Mike's next plan was to call and have a word with Philip.

He decided to take the bull by the horns and be completely up-front with him, not divulging Steve's name but saying that he was working for a client who wanted to put a stop to this stupid nonsense once and for all. Val had told Steve how upset Philip was and Steve had passed the word onto Mike, and knowing that he was St. Paul's right-hand man he felt it worth his while to make the trip to Newmarket.

He was not disappointed, Philip was beginning to reach the end of his tether and was glad for any help he could get. He was still one hundred percent loyal to St. Paul, but was banging his head against a brick wall, trying to make him see sense, he felt he was between the devil and the deep blue sea.

"I think what it is Mike, he knows what Boots and his mates are capable of, and he is really truly scared of a beating and will promise to do anything to try and ward them off, but they are not as soft as what he is and by making these promises is making it far worse for himself, when they cannot be kept."

Mike sent in his first report to Steve, and felt that he had made good headway in a couple of days. But there was more work to be done.

# Chapter Eighteen

A most beautiful sunrise, and a lovely sunny Saturday morning as two members of the band were preparing for their wedding that day. The band, of course, were going to play on this very special occasion, and form a guard of honour with their respective instruments when the couple left the Church.

The flower arrangements within the Church which the girl members of the band insisted upon doing personally as a wedding present were magnificent, blue and white, as the predominant colour of the band was blue. They had made their first visit to the New Spitalfields Market at Leyton to buy the flowers, and they were not disappointed. So much to choose from, and a shop where they could buy all the accessories they needed.

It was a truly happy occasion and everybody was in good spirits. David, the Best Man made an excellent speech, very funny, particularly about the antics of the Band.

When the opportunity arose, Philip came over to say how sad it was that Ken could not be with us on this very special day "He would have loved it Val" as well I knew.

"I know this is not exactly the right time to ask you a favour, but I wondered if you had a few minutes to spare one day, would you have a look in Ken's private cupboard at the Office, see if he left some scores of music there he was going to bring them with him to the Band Meeting, and as I know you have had so much to cope with recently, I wouldn't bother you with it until now, but only when you have a few minutes to spare."

"No problem, Philip, I will make a note and certainly find time during the week to have a look. I should have done something more about that cupboard before now. but there was such a build-up of work to get through it kept getting pushed to the end of the queue."

"A nice young man, name of Mike, called to see me one day last week and I must say I feel a lot better now that I know that someone professional is looking into this scam. He seemed to be with-it and was very open and above board with me, he knew of St. Paul's involvement, but didn't know the extent of it. I explained to him that I felt I must be loyal to St. Paul as I had worked for him for many years, and he has been kind to me in the past, but I didn't want to be mixed up with anything illegal. It has been a great burden to me over the last few months as you know."

"It's o.k. Philip, you mustn't worry about the call. I do know all about it".

"I felt that things were getting out of hand, and had a

word with Steve. He agreed to engage the Agent that he uses from time to time in his work as a Lawyer and that is the person who called on you, but I didn't know his plans at the time."

I was a bit worried about the cost, but Steve said he would help, and I am sure that if it reaches a satisfactory conclusion, the firm would be most willing to pay, so anything you can think of for him would be great."

Some of the worry-lines on Philip's face seemed to disappear instantly.

"He left me his card with his office and home number, and now I know who he is working for, I will rack my brains and let him know if I can remember anything of significance.

"Thanks Philip".

"Isn't this a wonderful wedding. This is the first time I have been really happy since that dreadful day last October."

I went to a little sandwich shop near where I worked, who did a roaring trade in the lunchtime and evening in the City with their very tasty sandwiches, took Kayla for a little walk, and returned to the office with the intention of making a start on the cupboard.

First of all I thought I would look through a few files again, see if I could find any connection to the present situation.

It did appear rather strange that a number of horses had died in the last two years as a result of laminitis. This proved nothing of course, but I felt that I had made a little headway and that I would have to investigate further, but I would get a book from the Library first and read up about laminitis, which was not going to solve the problem, but it may give me something more to think about. I was still puzzled about Philip being so upset, and wondered if he knew anything about why this was happening to these horses. I would have to make sure that I got into conversation with him the next time the band was playing at a function.

He was such a nice fella, and it could not be easy working for Saint Paul who was getting more eccentric by the day, and constantly upsetting Amanda by telling her what to do with the horses, although he knew nothing whatsoever about training. I understood, from Philip when we had our last chat, that the atmosphere at the Stable was very bad and the staff were becoming very distraught and threatening to leave.

I had still not made many inroads into the cupboard, but it was time to take Kayla home, I thought it not fair to her to have to do so much overtime there so soon in her new job.

She would let me know when the faxes were coming through by alerting me, and if there was a fire alarm practice she would alert me and then lie flat on the floor, and I would know it was the fire alarm. Although she is very protective, she had not been specifically trained to ward off any intruders, but I felt so much better about staying after the rest of the staff had gone home, having her with me.

I had not plucked up the courage to stay since the night someone was snooping in the office, but I felt it all worthwhile, at least I had a little something to go on, and I would have to play it by ear, but I couldn't wait to tell Steve.

He was quite cross with me…

"I told you Val, you must not stay there on your own in the evenings, it is too dangerous."

At least this meant that he was now taking the matter more seriously, and not putting it out of hand as he had in the beginning.

What fun it was travelling home together, everybody in the carriage fussing over Kayla, making my journey home so much more friendly. Normally everybody had their heads in the evening paper.

In the meantime Mike had been in contact with the market traders and had suggested to them that they play Chopper and his heavies at their own game. He said that he could round up quite a few Bouncers from Nightclubs, Dance Halls and other places that he knew who would be willing to stand up to the bullies next payment day, but of course they would want to be paid. The traders had a meeting and agreed that if they put £50.00 in each they would still be in pocket if no more demands were made in future, and that the so called insurance would be finished once and for all.

So on the following Friday a barricade was set up in the High Street and the bouncers were in position to confront the extortioners when they arrived.

Mike acted as spokesman, and told the mob if they wanted to act rough that they were ready for it, and would be in future if and when any more threats were made.

When they saw the size of the ex-wrestlers and heavy-weight boxers that had been specially picked for the job, they were in no hurry to hang around. They knew there and then that what had in the past been a very lucrative racket had now come to an end.

# Chapter Eighteen

A few days later Charlie got in touch with Mike and they agreed to meet that evening when the betting shop closed.

There was a double scotch waiting for Charlie when he arrived, and he knew he would be on a few quid from Mike who was always fair about any information received.

"There's a whisper going around Mike that Chopper's lost his steady income and is anxious to make it up elsewhere."

"That's not enough Charlie, I was the instigator of him losing it, so I want to know a lot more than that."

"Well, you know he and his mob sell drugs to the kids on the streets and at the rave-ups, but there is not enough in it for them, they are really only acting as distributors and they want to get hold of the stuff as it comes into the country, and buy direct, cut out the middle man so to speak."

"Sparks reckons he can guess who is supplying the transport, but he is keeping it to himself for the time being."

Charlie was paid well for this valuable tip, and was promised a much larger sum if he could follow it up with more information.

"I will be in touch Mike, always pleased to do business with you."

This gave Mike plenty to think about, but what the heck was it to do with horses feed being tampered with.

Mike always managed to put things together better when he got out his big blackboard and chalk and wrote out all the bits and pieces he had learned from the beginning.

The one name that came up with all the connections was Saint Paul. He was Amanda's father. He was Philip's employer. He knew Sparks, Boots and others from his prison visits, but how did all this tie up with the drug scene in East London, and Ken's accident?

Philip's request for the music scores gave me the inspiration to really get down to doing something about Ken's private cupboard. Steve had given me the key, and I had opened it once before and found the scrap-book of the band, but it was so personal, I could not bring myself, at the time, to delve further, but it had to be done, and I would have to brace myself to make a start.

It appeared that the cupboard was divided into three sections. The lower section had all the clippings and minutes of Band Meetings, music etc., and I eventually managed to find the scores of music Philip wanted.

The middle section related to the horses Ken owned and had an interest in.

The top section looked as though it housed all his personal papers. There was so much, how could anybody collect so much paperwork in such a short lifetime. It was obvious that it was going to take a considerable time to sort, but which section should I start on first. After consideration I decided to make a start on the band section. I would get a couple of empty boxes next time I visited the supermarket, put all the band stuff in them and pass them on to Philip and he could sort it all out at his leisure. That would clear at least one-third of the cupboard, but that was the easy

## Chapter Eighteen

part. I tossed a coin about what section I should start on next, and it came down heads, so I guessed that would mean the top section, all Ken's personal letters and things. I should have asked Steve to give me a hand with this, but he was so busy at this time, and he had his living to make, so I thought I would sort them into respective groups and give them to Steve and he could decide what to do with them.

I felt I was intruding into Ken's life, but it was a job that had to be done, and the best thing was to get on with it.

Eventually I came across a file with a letter from John.

I had heard Ken talk about John. Steve, John and Ken were in the same year at School and remained friends. I vaguely remembered Ken saying that John was a good Policeman. I read on.

*Hi Ken,*

*Glad to see you have been leading in some winners lately.*

*Ken, I need to ask your help. I know how much you despise the people getting the drugs to the youngsters today, and I think you may have a couple of contacts that we need some information about.*

*I prefer not to put too much on paper, so can we meet up some time you can come to my office, or I to your's or maybe a dinner one evening, whatever is most convenient for you.*

*Give me a ring and let me know.*

*Regards,*
*John*

Attached to this letter in the file were sheets of paper in Ken's handwriting, listing initials, dates and various routes from the Continent and other routes taken after leaving Dover and Harwich. This was a complete mystery to me, and I felt that I had to contact Steve, and let him sort it out.

I e-mailed him.

*I have found a file in Ken's cupboard – you should have it with some urgency – fax or e-mail me suitable time for handover. – Val.*

After inspecting the letter and file, Steve said he would give John a ring and arrange to meet up with him.

He went on..."It's all beginning to make sense now. Ken had made no secret of what he thought of the dealers taking advantage of the young people today. He always said that he, John and myself had made our way without the use of drugs and he could not understand why the youngsters needed them. Why they couldn't go out and have a good time in each others company without having to take ecstasy, coke or whatever.

This was a very strong point with Ken. He always appreciated the help we got in our early days, after our most unfortunate beginnings in life, with the early loss of parents, and yet it seemed today that it was the kids with loving parents and the way they were spoilt with too much money to spend who were experimenting with the various drugs that became available to them, and of course once hooked, that was it, the downroad that spiralled so quickly, and ruined their lives. I know Ken had put in a lot of background work to try and stem the flow a bit, but it was a heartbreaking job and what he was doing didn't seem to make the slightest indent.

"I feel sure this file has something to do with this, and I will get it to John as quickly as I can. Perhaps Ken was making more headway than we thought."

Mike paid Charlie another visit.

"Any news?"

"Well everyone knows of course that the so-called 'insurance' business has folded, thanks to you."

"What are they up to now?"

"Spark's says they are not content with just acting as distributors, there is not enough in it for them, and they want a bigger slice, they want to buy direct."

"They reckon they know one of the drivers bringing it in, but have yet to meet up with the person actually financing the import."

"They still think St. Paul knows something, and they are still threatening him, but it's way out of his league."

"He owns a transport company in Suffolk and they think his vehicles are being used, but if they are he would be the last to know, as the first sign of trouble he would spout, he is absolutely petrified of being beaten up."

## Chapter Eighteen

I paid a few visits to the Galloway Stable, and Simon was very pleased with Music Adored's work. He hoped to get her ready in time for Cheltenham. What a brilliant bit of news in an otherwise drab world.

# Chapter Nineteen

Mike's mobile was ringing – it always did at the most awkward times, it was Charlie.

"Meet me in the Duke's Head at 6:15p.m. and don't forget to bring your money."

Things looked promising if Charlie was expecting a good drop. Mike had a busy day ahead, much more work was being placed his way since his success with the market traders, but he desperately wanted a satisfactory conclusion to the assignment from Mr. Stevenson, it would open the door for him to get more work from Legal Offices, much more prestigious on the whole, and less likely to have to deal with the likes of the Chopper mob. Charlie was ready and waiting when Mike arrived.

"Let's get this over pronto Mike, I've got to be at the Dogs at 7 o'clock."

"Do you know of a chap called Speedie, he's a great get-away driver who Chopper uses from time to time.

He could compete in any rally, the very best in his line of work he is also an out-and-out bloody crook, he would nick anything, but he would never be able to make a living in the medical profession, can't stand the sight of blood, and passes out flat if anyone mentions that they are going to have a 'flu jab.

"Well, he is in a right old state at the moment, nerves completely shattered, drinking himself into oblivion, and mumbling in his drunken state that he didn't want to be mixed up with any killing. It's really got to him.

"I know they were returning from Portsmouth the day of the Kempton Meeting, they had been there to pick up a load, and Boots was moaning when they got back about the traffic, getting caught up with the racing crowd.

"Is this beginning to ring any bells?"

"It sure is, just stick with it Charlie, and with the right answers this could earn you a nice little bundle. Give me a ring just as soon as you have anything more about Speedie."

Mike sent his up-to-date report to Steve, and added that he felt that they had broken the barriers, and that quite soon all the necessary information would be flooding in.

Steve was over the moon.

If he was really true to himself he didn't hold out much hope when he first contacted Mike, he really did it to please Val, but Ken had been such a good mate, more like a brother really, that anything that could untangle the mystery would be a very good result.

## Chapter Nineteen

There was a fax in the machine when I got home.

*Val, don't want to raise your hopes too high, but Mick reckons he might be on to something.*

*Will contact you immediately if he comes up with anything else.*

John and Steve had arranged to meet for dinner. It was good for both of them, they had both taken the loss of Ken badly, like losing a brother, and they had not met up since the funeral.

Steve stowed the file into his briefcase, and hung on to it for dear life when he was on the tube.

After the first few preliminaries and settled with a drink, the file was handed over. John went on.

"I respect Ken for keeping this to himself, and as you can guess we don't spread the word around more than we can help, but I expect you realise by now, that Ken had been keeping his eyes and ears open for us for any information that would help in our undercover work.

"Briefly, there are far too many drugs getting into the country. We have put in a lot of work over the past year, and know about most of the vehicles being used, we could pull them up at any time, but we are letting them through and they are beginning to get a bit careless. We don't want to pick up bits and pieces here and there, we want to smash the whole operation from source to where it is being stashed, and this is the difficult part because we can follow a lorry from port, but we believe that the drugs are being transferred to other vehicles en route, and thus making it difficult for us to follow to the final destination, and this is part of what Ken has been trying to find out where the swap takes place, and tapping the file, he said I think this is going to be a big help."

"Do you think John that this had anything to do with Ken's accident?"

"No I don't think so Steve, Ken was very discreet and I don't think that they were aware of his involvement."

## Chapter Nineteen

Another call from Charlie –

"Mike we need to have another meet. I have had some expenses so bring plenty with you."

Mike was up to his eyes in work, but he particularly wanted to clear up the mystery regarding the motor accident, so decided to give it priority.

It was Wednesday, Charlie's afternoon off from the betting shop, so they met in the lunch-hour not at the Dukes Head as usual but at Charlie's request The Travellers Inn, a quiet little pub where they would not be recognised.

"I think I've got to the bottom of the riddle about the car going haywire – it's got to be worth a few quid."

"O.k. I said I would pay for good information, so give…"

"Well, I thought I would go to Speedie's local and accidentally bump into him if he were there, and he seems to be most of the time these days. Honestly Mike, he looks bloody awful. I gave it a bit of time, passed him on the way to the gents, and stopped to talk to him on the way back.

"Have a drink Speedie, I ain't seen you for a long time."

"I'll have a scotch Charlie, thanks."

"I got him a double and a chaser, and made my way over to his table "What's the problem Speedie, you don't look too good, got the 'flu?"

"No nothing like that, just a bit of bother."

"No work about for you at the moment?"

"There is, but I don't feel like doing it."

"I've never known you miss a trick before, it must be something bad if you're not interested."

"You know me Charlie, I don't mind turning over a place if it's going to be profitable, and it never hurts anyone, it's always somewhere where they would be well insured, and the only people hurt are the insurance companies, and they deserve it sometimes, what they charge for cover, and there is quite a bit of skill attached to it, how to get in, where to find the best stuff, and of course get out again without being caught, although it is getting harder with all the security lamps and alarms now, but in all my working life I've never hurt anyone, I couldn't, I haven't got the guts for it."

"Well what's happened to make you like this, no enthusiasm for work, that's got to be bad?"

"You know that rotten mob of Chopper's, all brawn and no brains, they asked me to drive them to Portsmouth in the lorry to pick up a load – I would never do it again, I've finished with them.

"Well on the way back we caught up with the race traffic coming out of Kempton, and when the Copper put his hand up to stop me to allow another line of traffic out from the Course, Boots started doing his nut, he recognised a car – a BMW – and the driver coming away from the car park into the stream of traffic making its way towards London. He was like a kid, 'get him Speedie knock him off the road'. Well of course I ignored that, and that made him mad.

"That's the bastard that tries to stop us getting to the raves and give the kids a good time."

Sticks piped in "How can you be so sure?"

"I'd know that number plate anywhere, I've seen it enough times."

"Well apparently they had been over to the Continent a couple of weeks previously, and somehow managed to pick up a couple of lasers. They had no idea at all how dangerous they can be I told them to put them away, but they wouldn't listen. Boots was going on about that man in Germany said you can cut through glass with these shall we try it – they were hanging out of the side windows. I begged them to be sensible, but I was on the outside lane of traffic and couldn't do much to stop them at the time, I had to keep going with the flow and as we passed the BMW they let fly with the lasers and from what I could see in my rear mirror they must have blinded the driver and he eventually crashed. I didn't hang about of course, but it's really got to me, to think that I've worked all my life without any rough stuff at all, and then just when I was thinking of retiring this should happen. I know I wasn't really responsible for killing that man, but I was the driver and I feel so guilty.

"Come on Speedie, have another scotch and chaser, make you feel better."

"I felt Mike I could be generous as I had learned so much."

"You've done very well Charlie, and I really appreciate it – you know I will be in touch if anything else crops up" at the same time stuffing a nicely folded number of notes in Charlie's top pocket.

The final report from Mike, together with his account was on Steve's desk when he arrived put together in the most professional way, and not at all as Speedie had described the incident, but ultimately reaching the same conclusion.

Steve was really pleased that Mike had dealt with this assignment so quickly and efficiently, and made a mental note to put the word around how good he was. He also thought that his fee was very reasonable and decided that he would stand this himself and not ask Val to approach the firm to contribute. He immediately faxed Val to arrange a dinner date, somewhere very special to put her at her ease, he was not quite sure how she was going to take it.

"This is very posh Steve, are we celebrating something?"

"I would hardly call it celebrating Val, but more like a conclusion to the mystery surrounding Ken's accident."

Steve could see the blood draining from Val's face "Look, it's what we set out to do, and now we know that it was not caused through any negligence on Ken's part, and that's what we will have to remember, because I don't think that there is any way we could prove what happened, and we shall in time just have to learn how to forgive".

He handed the report over to her so that she could digest it and let it sink in before they actually started on the dinner.

"In other words Steve, it was a couple of soppy sods, playing silly buggers, at an age when they should have known better. What they did could so easily have caused a multi-pile up with many people killed or seriously injured. They should of course, be punished, but as you say, there is no way we would be able to prove what happened, and I have always believed that what goes round comes around and they will pay for what they did in good time."

"I do feel that it was good that we both made the effort to prove that it was not down to Ken, and that now he will be able to rest in peace."

"Have you heard anything from John?"

"Not yet, but he was very pleased with the file we handed over, and said that Ken had been a great help to them with the very special assignment they have been working on for the past year to try and stem the flow of so many drugs entering the country.

It was all over the morning and evening papers.

*DAWN RAIDS WERE CARRIED OUT SIMULTANEOUSLY ON PREMISES AND HOUSES IN ESSEX, KENT EAST AND NORTH LONDON – THE DRUGS SEIZED WAS THE LARGEST HAUL EVER TO BE RECOVERED. SEVERAL MEN AND WOMEN WERE TAKEN INTO CUSTODY AND WILL BE CHARGED LATER IN THE DAY*

So the patient and intricate and very often exhausting work put in over the past year paid off.

It will never stop of course, but it will make it more difficult to get hold of for the time being, and hopefully some of the young people will be saved from a life of misery.

I wondered about the Police raids in East London, hoping particularly that Boots and Sticks, among others, were rounded up, and had a little giggle to myself at the thought of the Police pouncing on Sticks in bed in his long-johns, he is so thin they would think that they had hit on a ghost.

Steve faxed. "Have you seen the papers Val? You said what goes around comes around and it seems to be very true regarding the Chopper lot."

# Chapter Twenty

It was a drizzly, murky, overcast day when the last of the Kray family was to join his mother, father, elder brother Charlie and his twin at Chingford Mount Cemetery. A suitable day to finally see off the last member of a family who put so much fear and unhappiness into other people's lives.

It still baffles many police and journalists that so much interest in the Krays still prevails today. Like all the old East-end families their funerals are a big thing – must have masses of flowers and eventually a lavish headstone for the grave which would cost a fortune, when for some people there was hardly enough money to put food on the table. As always, when a well-know person, notorious or otherwise, dies and it is reported by the various media, everyone jumps on the bandwagon to give various interviews, and write articles about how well they knew the deceased.

A lot of rubbish was said and written about the Krays,

how kind they had been to the poor people of Bethnal Green. They only gave away what they had stolen from other people, and that was mostly to get publicity.

They were fanatical about being photographed with celebrities, and would make special journeys to clubs and restaurants in the West End with the sole purpose of standing behind the chair when they spotted a celebrity, photographer at the ready to snap them as though they were best friends.

George Raft, an excellent American film-star, and the most fantastic dancer ever, he could rumba, tango, samba etc. to perfection, was quickly pounced on by the Krays when he visited this country and was eventually deported, many people believed it was because he became involved with the Krays.

The service had to be at the Bethnal Green Church they only used for Hatch, Match and Dispatch. Why they always wanted such a service, when all their lives they showed not the slightest sign of Christianity, was always a mystery, it was just because it was an East-end tradition, and right until the last they had to put on a big show.

The glass hearse, driven by two beautiful horses, all dressed up for the occasion, was to parade through the heart of East London to the final resting place at the top of Chingford Mount. It was to take one hour to travel to the Mount, but eventually took two.

Every road and street was crowded by onlookers, and every hoodlum, past and present, all dressed in the same black garb, lined the route.

Those notably missing were Chopper, Sticks, Boots and the rest of their mob.

# Chapter Twentyone

Philip called with a letter from the Band asking me to accept the Vice-Presidency.

"You know, of course, Val that Amanda has been arrested, she is being held at a Police Station somewhere in Suffolk?"

"Hang on a minute Philip, I don't think I am lipreading you properly – did you say Amanda has been arrested?"

"Yes, yesterday morning – poor old St. Paul is taking it very badly."

"But whatever for, what has she done?"

"Amanda was the person ultimately responsible for the whole operation of getting the drugs from the original source into this country. She was financing the operation, and working with St. Paul's old mate, Dodgy Duggie."

"It was Duggie who persuaded Paul to invest in the transport company in the first place, and he then ran it for him."

"I can't believe this, she always thought she was so superior to everybody else."

"The Police knew that she was always up early and at her Yard by six a.m. every morning, so they raided her house at four a.m. and got her out of bed."

"But why, Philip, why would she get involved with something like this, surely not for the money – she can't be hard-up, and even if she was, St. Paul would see her alright."

"I think, she just must be in charge, she must be the boss, and the thrill of training winners was wearing a bit thin and she felt the need to be involved with something a bit more risky – it's the only thing I can think of, she enjoys taking a chance, and this was a very big one."

"I thought it was St Paul who was being threatened."

"It was, they were getting at him to get to her, but she is made of sterner stuff than her father, and could well hold her own, although she was very upset when the horses started dying, but in some strange way it seemed to egg her on to take every greater risks."

"You know it has been a big worry to me over the past few months, although I didn't know the full extent of it. Amanda kept her dodgy dealings with Duggie well away from my ears, although I could see something was going on. Her husband is going bonkers, I think it's the end of his hopes of a Knighthood."

"Do you think Philip, that Ken had some idea about Amanda's involvement with all this. You know he took his horses away from her Yard and put them with Simon Galloway. I never really understood why, but I can see now if Ken thought that she had any tie-up with

## Chapter Twentyone

drugs at all he wouldn't hesitate, he held such strong views about the people getting drugs on the streets to teenagers and young people who were about to begin a very important time in their life, which could so easily be ruined."

"Poor old Amanda, she won't half notice the difference in Holloway if that's where they put her, she won't be able to put on her airs and graces there, they will knock her down a peg or two."

# Chapter Twentytwo

I went to early Mass. I particularly wanted to have a word with the Holy Mother, thank her for helping me to put the record straight about Ken, his good name fully restored regarding the accident. John had told Steve how helpful Ken had been about meticulously keeping records of the transport used to distribute, particularly in East London, and this had played an important part in the final roundup. So I wanted to thank the Holy Mother for her Prayers to bring this distasteful and horrifying trade to an end, at least for the time being.

I flew out of Church and rushed to meet the three coaches that were going to make their way to Cheltenham that day. The coach for the Band, who had been practising 'Congratulations' should Music Adored live up to expectations.

The coach for the Signing Choir who were to sign with the Band wearing West Ham colours should the opportunity arise, and of course the coach for the boys and girls at the office, who had been so loyal and faithful to the Boss's horse.